Rainer H. Müller, Wolfgang Mehnert

Particle and Surface Characterisation Methods

Particle and Surface Characterisation Methods

edited by

Prof. Dr. rer. nat. Rainer H. Müller
Professor of Pharmaceutics & Biopharmaceutics
Free University of Berlin

and

Dr. rer. nat. Wolfgang Mehnert
Senior Lecturer
Department of Pharmaceutics and Biopharmaceutics
Free University of Berlin

with the participation of

Dr. Gesine E. Hildebrand
Lecturer
Department of Pharmaceutics, Biopharmaceutics & Biotechnology
Free University of Berlin

The book chapters are based on the invited lectures
presented by the authors at the Colloidal Drug Carriers
– cdc – 2nd Expert Meeting
Mainz (Germany), 6 March 1997
(main organizers: Prof. Dr. R. H. Müller
and Dr. W. Mehnert, FU Berlin)

140 Figures and 29 Tables

medpharm Scientific Publishers Stuttgart 1997

The use of general descriptive names, tradenames, trademarks, etc. in a chapter, even if not specifically identified, does not imply that these names are not protected by the relevant laws and regulations.

Prof. Dr. Rainer H. Müller
Department of Pharmaceutics, Biopharmaceutics & Biotechnology
Free University of Berlin
Kelchstr. 31
D-12169 Berlin

Dr. Wolfgang Mehnert
Department of Pharmaceutics, Biopharmaceutics & Biotechnology
Free University of Berlin
Kelchstr. 31
D-12169 Berlin

Die Deutsche Bibliothek – CIP-Einheitsaufnahme
Particle and surface characterisation methods:
based on the invited lectures presented by the authors at the Colloidal drug carriers - CDC, 2nd expert meeting, Mainz (Germany), 6 March 1997 ; 29 tab. / ed. by Rainer H. Müller and Wolfgang Mehnert, with the participation of Gesine E. Hildebrand. - Stuttgart : medpharm Scientific Publ., 1997
ISBN 3-88763-057-2

All rights reserved. No part of this publication may be translated, reproduced, stored in a retrieval system, or transmitted, in any form or by any means, electronic, mechanical, photocopying, microfilming, recording or otherwise, without permission in writing from the publisher.

© 1997 medpharm GmbH Scientific Publishers, Birkenwaldstraße 44, D-70191 Stuttgart
Printed in Germany

Druck: Karl Hofmann, Schorndorf
Umschlaggestaltung: Atelier Schäfer, Esslingen

Preface

The characterisation of surfaces is an important tool for quality control (e.g. reproducibility of batches) and also essential prerequisite for a better understanding of the interactions of surfaces with their environment, i.e. liquid but also gaseous media. This includes flat surfaces (e.g. films, surfaces in medical devices, polymeric implants in the body etc.) as well as curved surfaces of particulates (e.g. powders, suspensions, nanoparticles, liposomes and emulsion droplets).

Surface area and surface properties such as the nature of functional groups and surface hydrophobicity determine the interactions with the dispersion medium, e.g. nature and quantity of adsorbed solutes and gas. Good examples are pharmaceutical products. The biocompatibility of medical implants depends very much on their surface properties. The organ distribution of intravenously injected drug carriers such as liposomes, nanoparticles and emulsions is a function of the physicochemical surface properties which determine the composition of the adsorption pattern of blood proteins. These proteins mediate the adherence and uptake by different cells (concenpt of differential adsorption). Apart from the qualitative and quantitative composition of the protein adsorption pattern, the conformation of the proteins plays also an important role (e.g. exposure of binding moieties).

In this book, a full range of different characterisation methods is presented. The aim is to provide an overview to the reader allowing him to identify the method of choice for his special analytical problem. The chapters contain information about the basic theory, possibilities but also limitations in resolution and application of the various methods.

The methods include measurement of particle size using laser light scattering techniques. It also presents the latest development - the PIDS technology. A central point - the validation of software and hardware - has also been considered. Determination of surface area is described focussing on modern BET instruments.

FTIR techniques are included for chemical analysis of the surface. A challenging task is the quantification of surface hydrophobicity. How to measure it? One approach is the measurement of contact angles on dried flat surfaces (in case of polymeric particles: e.g. compressed tablet or casted film after dissolution of the particles in an organic solvent). However, this has the serious disadvantage that the properties of such films can differ very much from the properties of a surface in cocntact to a dispersion medium such as water. The surface might be hydrated or solvated affecting strongly its hydrophobicity. In addition, liposomes and emulsion droplets are not accessible. In this book dye adsorption methods based on Rose Bengal and Hydrophobic Interaction Chromatography (HIC) are presented allowing the measurement in the original aqueous dispersion medium or at least a medium similar in composition (in case of HIC).

Interaction of surfaces with proteins in the surrounding medium are a central question in many different areas. Classical examples are medical devices, implants and parenterally injected particles (diagnostics and also drug carriers such as liposomes and emulsions). Two-Dimensional Polyacrylamide Gel Electrophoresis (2-D PAGE) is presented for semi-quantitative analysis of adsorbed proteins. The technique has been especially modified to allow processing of particles. In addition, to approach the problem of conformation, Circular Dichroism is included.

Berlin, 31 January 1997

        Prof. Dr. Rainer H. Müller   Dr. Wolfgang Mehnert

Content

1 Multiangle Photon Correlation Spectroscopy in Particle Characterization (R. Xu) 1

1.1 Introduction 1
1.2 Theoretical Background 1
1.3 Multiangle PCS Measurement 9
1.4 Acknowledgement 17
1.5 References 17

2 Possibilities and Limitations of Laser Light Scattering Techniques for Particle Size Analysis (I. Zimmermann) 19

2.1 Scattering by a Single Particle 19
2.2 Scattering by Collective Samples 22
2.3 Discussion 26
2.4 References 26

3 Improvements in Particle Size Analysis Using Light Scattering (R. Xu) 27

3.1 Introduction 27
3.2 Angular Light Scattering Intensity 29
3.3 Improvements in Submicron Size Analysis 41
3.4 PIDS Instrumentation and Data Analysis 50
3.5 Summary 54
3.6 Appendix 55
3.7 References 55

4 Particle Sizing in the Sub-micron Range by Laser Diffraction (M. Wedd) 57

4.1 Introduction 57
4.2 What is Refractive Index? 63
4.3 Data Inversion 64
4.4 Calculation of Volumetric Concentrations and Specific Gravities ... 65
4.5 Practical Results 66
4.6 References 67

5 A Guide to the Validation of Laser Diffraction-based Particle Size Analysers (M. Wedd) 69

5.1 Introduction 69
5.2 The Regulatory Environment Structure 69

5.3 The Concept of Validation 73

6 Confocal Laser Scanning Microscopy and its Application in Liposomal Research (T. Möller) 85

- 6.1 Summary 85
- 6.2 Introduction 85
- 6.3 Principles of Confocal Microscopy 86
- 6.4 Resolution 88
- 6.5 Lasers 91
- 6.6 Dyes 92
- 6.7 Limits of Confocal Microscopy 92
- 6.8 Applications of Confocal Microscopy 94
- 6.9 Further Readings 98
- 6.10 References 98

7 Introduction to Atomic Force Microscopy and its Application to the Study of Lipid Nanoparticles (E. zur Mühlen, H. Niehus) 99

- 7.1 Introduction - Traditional Microscopy Techniques 99
- 7.2 Introduction - Scanning Probe Microscopy 101
- 7.3 Imaging Mechanism of STM 102
- 7.4 Forces Acting between Probe and Sample 105
- 7.5 Contact-AFM: Topography Imaging 109
- 7.6 Contact-AFM: Lateral Force Microscopy (LFM) 115
- 7.7 Contact-AFM: Force-Distance Curves 117
- 7.8 Contact-AFM: Scanning Force Spectroscopy (SFS) 121
- 7.9 Non-Contact AFM 122
- 7.10 Conclusion 126
- 7.11 References 127

8 FTIR Techniques for Surface Characterization (A. Büchtemann, R. Dietel) 129

- 8.1 Introduction 129
- 8.2 Basic Information on Infrared Spectroscopy 130
- 8.3 Transmission Spectroscopy 133
- 8.4 Reflection Spectroscopy 138
- 8.5 References 154

9 Circular Dichroism (CD) for the Analysis of Dissolved and Adsorbed Proteins (H. Hermel) 159

- 9.1 General Remarks 159
- 9.2 Origin of Absorption and Dispersion 161

9.3 Poly-L-lysine CD, a Standard for the Secondary Structure 162
9.4 Problems of CD-investigations of Adsorbed Proteins 162
9.5 In situ Measurements of Adsorbed Protein Layers at the Fluid Interface...... 164
9.6 References.. 168

10 The Surface Structure of Lipid Drug Carriers - Influence on Carrier-Cell Interaction (L. Bergelson, A. Domb) 169

10.1 The Heterogenity of the Lipid Particle Surface 169
10.2 Methods to Study Lipid Surface Structure 171
10.3 Influence of Proteins and Peptides on the Surface Structure of Lipid Drug Carriers 177
10.4 Influence of Drug Loading on the Carrier Surface 178
10.5 References 181

11 Surface Area Analysis of Finely Divided and Porous Solids by Gas Adsorption Measurements (T. Schoofs) 185

11.1 Introduction 185
11.2 Fundamental Understanding of Surface Area 185
11.3 Characterisation Methods 187
11.4 Physical Adsorption of Gases 189
11.5 The Adsorption Process 190
11.6 Determination of Surface Area 193
11.7 Practical Measurement 195
11.8 References 197

12 Determinations of Surface Area in Comparison (H. Winter) 199

12.1 Sample Preparation - Single-point- and B.E.T. - Multi-point Measurement 199
12.2 Preparation of the Samples 200
12.3 Choosing of the Adsorbate Gas 201
12.4 Conclusions 201
12.5 Description of the Measurement Results 202
12.6 Test 18 and 19 206
12.7 Summary 207
12.8 References 208

13 The Surface Area of Magnesium Stearate: An Example for a Complex Analysis Task (F. Metz) 209

13.1 Introduction 209
13.2 Considerations for Small Surface Areas 209
13.3 Sample Preparation Considerations 211

13.4 Analysis of Commercial Magnesium Stearates 212
13.5 Conclusion 213

14 Surface Hydrophobicity - Determination by Rose Bengal (RB) Adsorption Methods (R.H. Müller) ... 215

14.1 Importance of Hydrophobicity of Particles and Surfaces for Physical and Biological Interactions 215
14.2 The Basic Problem of Measuring Surface Hydrophobicity 217
14.3 Rose Bengal Binding Methods 218
14.4 Summary 226
14.5 References 227

15 Theory and Set Up of Hydrophobic Interaction Chromatography (HIC) (K. Thode, R.H. Müller) ... 229

15.1 Theory of HIC 229
15.2 Running HIC 231
15.3 Running Mini - HIC 232
15.4 References 233

16 Hydrophobic Interaction Chromatography (HIC) for Determination of the Surface Hydrophobicity of Particulates (R.H. Müller) ... 235

16.1 General Applications of HIC 235
16.2 Experimental 235
16.3 Characterization of Polystyrene Latex Particles with Different Functional Surface Groups 237
16.4 HIC of Polymeric Particles Surface-modified by Polymer Adsorption 241
16.5 Application of HIC to Differently Sized Nanoparticles, Emulsions and Liposomes 246
16.6 Summary 248
16.7 References 249

17 HIC of Iron Oxide Dispersions Stabilized by Different Macromolecules (K. Thode, M. Kresse, R.H. Müller) ... 251

17.1 Introduction 251
17.2 Choice of the Appropriate Wavelength 252
17.3 Running Mini-HIC of Iron Oxide Dispersions 255
17.4 Conclusions 259

18 Two-Dimensional Polyacrylamide Gel Electrophoresis (2-D PAGE) for the Analysis of Protein Adsorption onto Surfaces (T. Blunk) 261

   18.1 Why 2-D PAGE for the Analysis of Protein Adsorption? 261
   18.2 How to Do the Analysis? 262
   18.3 What to Expect from 2-D PAGE? - A Study of Kinetics of Protein Adsorption as an Example 268
   18.4 Conclusion 272
   18.5 References 273
   18.6 Abbreviations 275

19 Index 277

1 Multiangle Photon Correlation Spectroscopy in Particle Characterization

Renliang Xu, Coulter Corporation, Miami, Florida, USA

1.1 Introduction

During the past two decades, photon correlation spectroscopy (PCS) has become a mature and popular technology for probing the diffusion of particulate materials in solution or in suspension. By determining the rate of diffusion, or the diffusion coefficient, information regarding the size of solid particles or the conformation of macromolecular chains can be obtained. Commercial PCS instruments have evolved from heavy and bulky optical setups with furniture-sized electronics and correlators, to miniature setups utilizing fiber optics and chip correlators based on digital signal processing (DSP). On-line or in-line particle characterization using PCS has become viable, providing a fast and non-invasive means to monitor or control product quality in many manufacturing processes. An international standard for using PCS to characterize spherical particles has recently been established by ISO [1]. In this chapter, the principle of PCS will be summarized in the second section. The features of multiangle PCS measurement will be the topic of the third section, followed by a short summary.

1.2 Theoretical Background

Photon correlation spectroscopy, also termed dynamic light scattering (DLS) or quasi-elastic light scattering (QELS), is a branch of light scattering that has often been used in different disciplines of science [2]. In a PCS experiment, the fluctuation of scattered light from scatterers in a liquid medium is recorded and analyzed, either in the frequency domain or in the correlation delay time domain depending on the application. The scatterers can be any particulates that have a different refractive index than that of the medium, and are stable during the measurement. Typical scatterers are solid particles in suspension, such as metal oxide, mineral debris and latex particles; soft particles in suspension, such as vesicles and micelles; and macromolecule chains in solution, such as synthetic polymers and biomaterials. For the sake of simplicity, we will call all these scatterers particles in the present text. The detected scattering may be directly from individual particles (single scattering), or from multiple scattering in the cases of concentrated solutions or suspensions. Fluctuations in the measured scattered-light intensity at a given scattering angle arise because Brownian motion of the particles continually rearranges the configuration of particles in the scattering volume. The relative positions of the particles in the scattering volume at any instant determines the degree of constructive or destructive interference of scattered light at the detector. Since the diffusion rate, or velocity, of the particles is determined by their size (given that fluid viscosity and temperature are known or constant), information about size is contained in the rate of fluctuation of scattered light intensity. The lower particle-size limit for measurement is determined by the scattering intensity and the experimental noise. The scattering fluctuations to be measured must be greater than the experimental noise from sources including environment disturbances, temperature fluctuations, and electronic noise in order to obtain unbiased results. The upper particle-size limit for measurement is determined primarily by the sedimentation limit. Particles must be stably suspended, because sedimenting particles leave the scattering volume, and undergo directed motion which contributes complicating information to the scattered-light fluctuations. Practically, the upper size limit for a PCS experiment is about a few µm depending on material density and dispersant (solvent) viscosity, and the lower size limit is about a few nm depending on the refractive index contrast between particles and dispersant (solvent).

Depending on the application and experimental setup, there are several PCS techniques, such as diffusive wave spectroscopy and transient scattering spectroscopy, and several detecting schemes, such as self-beating, homodyne, and heterodyne [3]. In this chapter, we will limit our discussion to the most common PCS application: scattering from particles in dilute suspension (or dilute solution) in the self-beating mode, in which particles move and scatter independently without influence from other particles in the sample. Through the motion of particles in the medium, information on particle dimension (or conformation for chains) can be obtained. This technique is one of the most popular non-invasive, reproducible, physico-chemical methods used in particle characterization and polymer and colloid studies. To illustrate the reproducibility of the method, Fig. 1-1 and Fig. 1-2 present the result from repeat PCS measurements of a monodisperse polystyrene latex of nominal diameter 300 nm dispersed in n-butanol. The measurements were performed using four PCS instruments (Coulter N4 Plus) with 20 repeating measurements for each instrument.

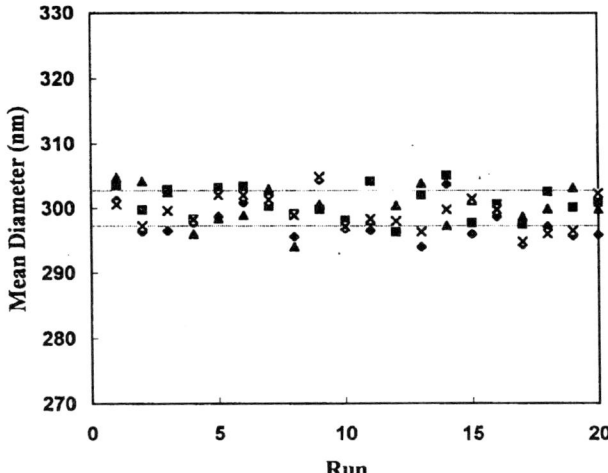

Fig. 1-1: The mean diameters of 300 nm polystyrene latex particles dispersed in n-Butanol recovered from PCS measurements using four Coulter N4 Plus instruments at a scattering angle of 90 degrees. The cumulants method is used to obtain the mean diameters. 20 consecutive runs were performed on each instrument. The overall standard deviation of the experiment is 0.9% (two dotted lines), which is much better than that specified in the ISO standard (<5%) [1]. The sample has been kept in a sealed cell for five years.

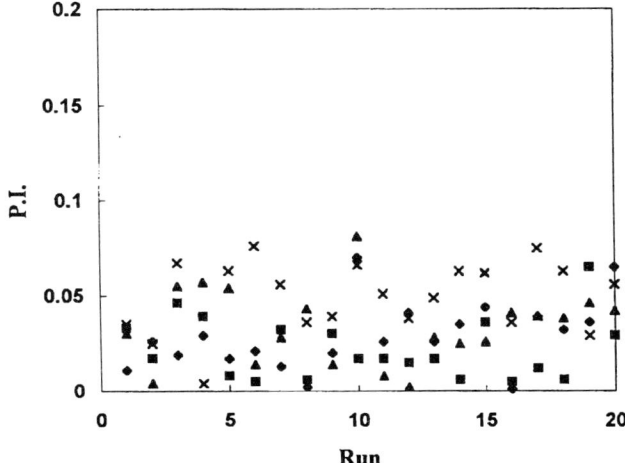

Fig. 1-2: The polydispersity index (P.I.) values of the 300 nm polystyrene latex sample obtained from the same experiment as in Fig. 1-1.

Fig. 1-3 is a schematic of a typical optical setup for a multiple-angle PCS instrument. The light from a He-Ne laser or an argon ion laser is commonly used as the light source because of high stability and long coherence length. Recently, diode lasers have been more commonly employed as the light source because of the advantages of low cost, miniaturization, ruggedness and connectability with fiber optics. The sample cell can be either a simple square or round cuvette, or a more sophisticated type with better temperature stability and higher optical quality. A typical sample volume is a few milliliters or less. Particles in the scattering volume, which is the cross section between the light beam and the detecting optics, scatter light in all directions. The scattered light is collected at a scattering angle θ, either by an aperture-lens setup or by an optical fiber, and then detected by a photo-electric detector [3]. Conventionally, a fast-response and low-noise photomultiplier tube (PMT) has been used as the detector. Recently, however, many examples have shown that an avalanche photodiode (APD) can be successfully used as the detector [4]. The electric pulses from the photo-electric detector are amplified, stretched and fed into an autocorrelator for computing the intensity autocorrelation function (ACF) $g^{(2)}(\tau,\theta)$, which is the primary data from a PCS measurement.

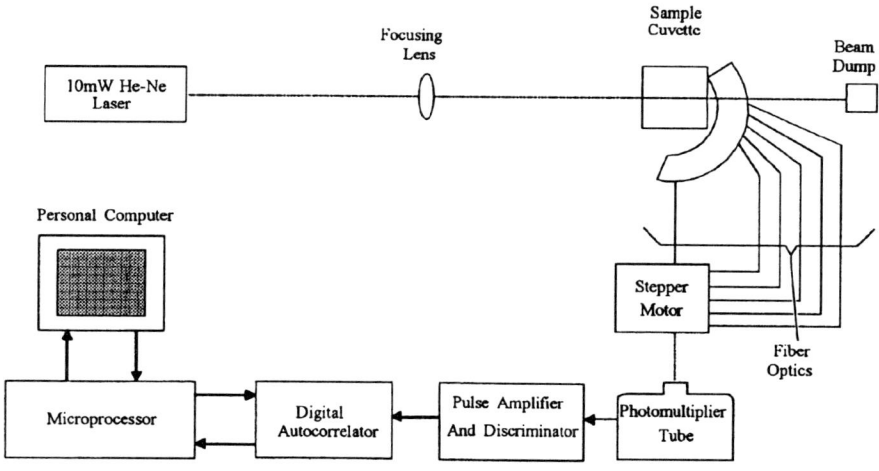

Fig. 1-3:   Schematic of a typical multiangle PCS instrument.

The normalized intensity ACF $g^{(2)}(\tau,\theta)$ is computed according to:

$$g^{(2)}(\tau,\theta) = \frac{<I(t,\theta)I(t+\tau,\theta)>}{<I(t,\theta)>^2} \qquad \text{Eq. 1-1}$$

In Eq. 1-1, $I(t,\theta)$ is the scattering intensity collected at a scattering angle θ and at the time t. τ is the correlation delay time. The average is over the entire measurement time. For an optical field that obeys Gaussian statistics, $g^{(2)}(\tau,\theta)$ is related to the electric field ACF $g^{(1)}(\tau,\theta)$ by the Siegert relation:

$$g^{(2)}(\tau,\theta) = 1 + \beta |g^{(1)}(\tau,\theta)|^2 \qquad \text{Eq. 1-2}$$

where β is an instrument efficiency coefficient that has a positive value of unity (in the ideal situation) or less. For a dilute suspension (or solution) of monodisperse non-interacting particles, $g^{(1)}(\tau,\theta)$ is an exponential decay function with a characteristic decay constant Γ:

$$|g^{(1)}(\tau,\theta)| = \exp(-\Gamma(\theta)\tau) \qquad \text{Eq. 1-3}$$

When there is more than one characteristic decay from the sample, e.g., the sample is polydisperse, the addition of the scattered light from individual particles makes $g^{(1)}(\tau,\theta)$ a function of multiple exponentials:

$$|g^{(1)}(\tau,\theta)| = \sum F_i(\Gamma_i,\theta)\exp(-\Gamma_i(\theta)\tau) = \int_{\Gamma_{min}}^{\Gamma_{max}} G(\Gamma,\theta)\exp(-\Gamma(\theta)\tau)d\tau \qquad \text{Eq. 1-4}$$

In Eq. 1-4, i represents a group of particles with the population F_i that has the same characteristic decay constant Γ_i. $F_i(\Gamma_i,\theta)$ or $G(\Gamma,\theta)$ are the discrete distribution and the continuous distribution, respectively, to be extracted from the ACF. Γ is related to the translational and rotational motions of the particles in the given medium. For a macromolecule chain, Γ is mainly determined by the chain length and conformation in the medium. For a soft particle, Γ relates to the softness, the flexibility, the dimension and shape of the particle. For a solid particle, Γ is a characteristic of the shape and the dimension. For solid spherical particles, since there is no detectable rotational motion, Γ is directly related to the diameter d through the well known Stokes-Einstein equation:

$$\Gamma = K^2 D_T = \left(\frac{4\pi n \sin(\theta/2)}{\lambda_0}\right)^2 \frac{k_B T}{3\pi\eta_0 d} \qquad \text{Eq. 1-5}$$

with K, D_T, n, λ_0, k_B, T and η_0 being the magnitude of the scattering vector, the translational diffusion coefficient, the refractive index of the medium, the wavelength of light in vacuo, the Boltzmann constant, the absolute temperature, and the solvent viscosity, respectively.

Once $G(\Gamma,\theta)$ is obtained, one is be able to obtain the dimension of the particles, which often is the primary interest of a PCS experiment, when combined with other known information of the sample. Thus, extracting $G(\Gamma,\theta)$ in Eq.4 from the measured $g^{(1)}(\tau,\theta)$ is the key step in the data retrieving process.

Generally speaking, there exist three routes for obtaining $G(\Gamma,\theta)$ from the ACF. The first one is to obtain the mean value of Γ and the breadth of the distribution, characterized by the polydispersity index (P.I.). This type of data retrieving procedure disregards the details of $G(\Gamma,\theta)$ but provides fairly good mean values of Γ. There are several algorithms in this category. The cumulants method has been proven to be the best one by a broad range of practice [5]. The cumulants method is not very sensitive to the data quality and experimental noise. The current ISO Standard recommends only the cumulants method in PCS measurement of particle size. In the cumulants method, multiple exponential decays are compressed into a single exponential with the exponent expanded into a polynomial of the delay time τ:

$$\text{Ln}\left|g^{(1)}(\tau,\theta)\right| = -<\Gamma(\theta)>\tau + \frac{\mu_2(\theta)}{2}\tau^2 + \dots \qquad \text{Eq. 1-6}$$

where $<\Gamma>$ is the mean decay constant defined as:

$$<\Gamma(\theta)> = \int_{\Gamma_{min}}^{\Gamma_{max}} \Gamma(\theta) G(\Gamma,\theta) d\Gamma \qquad \text{Eq. 1-7}$$

The P.I., a measure of the breadth of the distribution, is defined as

$$P.I. = \frac{\mu_2(\theta)}{<\Gamma(\theta)>^2} = \frac{1}{<\Gamma(\theta)>^2} \int_{\Gamma_{min}}^{\Gamma_{max}} (\Gamma(\theta) - <\Gamma(\theta)>)^2 G(\Gamma,\theta) d\Gamma \qquad \text{Eq. 1-8}$$

Obtaining $<\Gamma>$ and the P.I. from the ACF using the cumulants method can be accomplished by a least-squares fitting of the ACF (or the logarithm of the ACF with a proper statistical weighting for each data point).

The second route to obtain the Γ distribution is by inserting a known analytical function with a few variables into Eq.4. Such a function usually has two variables, one is related to the mean value and the other one is related to the distribution broadness. These two variables are fitted out through the regression of Eq. 1-4 The following functions are commonly used in particle characterization.

Differential Distribution Form	Mean	Mode
Normal Distribution (Gaussian Distribution) $$p(x) = \frac{1}{\sqrt{2\pi}\sigma} \exp(-\frac{(x-a)^2}{2\sigma^2})$$	a	a
LogNormal Distribution $$p(x) = \frac{1}{x} \exp(-\frac{1}{2\ln(\sigma+1)} (\ln \frac{x\sqrt{\sigma+1}}{a})^2)$$	a	$a/(\sigma+1)^{3/2}$
Rosin Rammler Sperling Bennet Distribution $$p(x) = x^{n-1} \exp(-\sigma x^n)$$	$\dfrac{\Gamma(\frac{1}{n}+1)}{n^2\sigma^{1.5}}$	$\sqrt[n]{\dfrac{n-1}{\sigma n}}$
Schulz Zimm Distribution $$p(x) = \left(\frac{m}{a}\right)^{m+1} \frac{x^m}{\Gamma(m+1)} \exp(-\frac{mx}{a})$$	$\dfrac{a(m+1)}{m}$	a

Tab. 1-1: Common Distribution Functions

In recent years, with development of the third route for obtaining the Γ distribution, the Laplace inversion, the functional form is used less because of the arbitrariness in choosing a proper function. The functional forms are now used primarily for theoretical modeling and simulation. Eq.4 is a Laplace equation if Γ_{min} and Γ_{max} are extended to zero and infinity, respectively. To resolve $G(\Gamma,\theta)$ from the known $g^{(1)}(\tau,\theta)$ requires a Laplace inversion. Because there is noise from various sources in the measured $g^{(1)}(\tau,\theta)$, and because the measurement has a limited bandwidth, the Laplace inversion is ill-posed. Through intensive research and development of information retrieval procedures to obtain $G(\Gamma,\theta)$ from PCS measurements, both in mathematical formulation and in practical verification, several feasible and reliable algorithms have emerged : namely, regularized non-negative least-squares, singular value analysis, exponential sampling, and the maximum entropy method. In particular, the computation routine CONTIN by Provencher utilizing a regularized non-negative least-squares technique has won wide acceptance and has been implemented as the main data retrieving algorithm for PCS in many research laboratories and several commercial instruments throughout the world [6]. CONTIN is often used as the reference for evaluating the performance of different algorithms. If an ACF is measured with careful sample preparation, temperature stabilization, and extended data acquisition time to avoid noise from sample inhomogeneity, trace "dust" particles and temperature fluctuations, the information in the ACF can be well retrieved using a Laplace inversion technique. However, due to the nature of the technology, the resolution for the distribution retrievable from a PCS measurement is low. Generally speaking, a peak resolution of no less than two in the Γ distribution can be achieved for ideal samples such as mixtures of monodisperse spheres. In other words, if the subpopulations in the sample have their peak ratio $\Gamma_{peak1}/\Gamma_{peak2} \geq 2$, these two subpopulations can be resolved. Of course, the above statement is subject to the width and distribution shape of each subpopulation. Practically, a peak resolution of three is not difficult to achieve. There are many examples in the literature for the effectiveness of the Laplace inversion in retrieving multimodal distribution in a broad Γ range, typically in 2 to 3 decades, both by computer simulated ACFs or real sample measurements [for example, 7]. Due to the nature of the exponentially decaying ACF, the inversion technique works in a logarithmically scaled space. The output $F_i(\Gamma_i,\theta)$ or $G(\Gamma,\theta)$ from such inversions are also logarithmically spaced where $F_i(\Gamma_i,\theta)=G(\Gamma_i,\theta)\Gamma_i$; consequently, an equal resolution through the distribution is only true for $\log(\Gamma_i)$, not for Γ_i. Since Γ is inversely proportional to d, the size distribution obtained also has the equal resolution for $\log(d_i)$, not for d_i. The detail descriptions of the Laplace inversion algorithms are beyond the scope of the present text. Fig. 1-4 gives an example of the particle size distribution retrieved from a bimodal polystyrene latex (PSL) mixture measured by Coulter N4 Plus using the CONTIN algorithm.

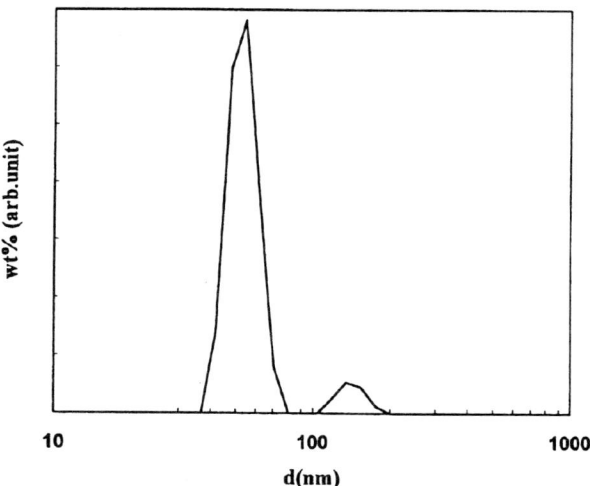

Fig. 1-4: A bimodal mixture of polystyrene latex particle in aqueous suspension retrieved from PCS measurement using the Coulter N4 Plus instrument. The latex particles from Duke Scientific have the nominal values of 50 nm and 155 nm, respectively.

1.3 Multiangle PCS Measurement

In PCS measurements, the information regarding particle motion is obtained from the scattering intensity fluctuations. For monodisperse samples or samples having narrow size distributions, if motions other than translation are discounted, there is only one characteristic decay constant and the scattering from all particles will not be received discriminatorily. Once a proper delay time is chosen for the ACF at a given scattering angle, the size information can be retrieved with little bias. Thus, many PCS instruments are only equipped with single angle detection, usually at the right angle (90°), and consequently experiments are only performed at one angle. However, for a polydisperse sample, to obtain an unbiased result for the mean size or the size distribution is not an easy task. A measurement at one scattering angle is often inadequate because particles of different sizes have different scattering intensities and different decay constants at a given scattering angle. Measurement at more than one angle, a multi-angle measurement, provides several advantages and is often necessary in order to obtain correct results.

The biggest difficulty in using PCS to characterize a polydisperse sample arises from the fact that the angular scattering intensity patterns per unit volume for particles of different sizes (mass) are quite different due to intraparticle destructive interference. For simplicity, let us just discuss the case of solid spheres. The angular scattering pattern for solid spheres can be described by a type of spherical Bessel function following the Mie theory [8]. These patterns are oscillatory. They all have a maximum at the zero degree scattering angle followed by a sequence of minima and maxima with decreasing intensity. Fig. 1-5 shows three angular scattering patterns of unit volume scattering from 2 µm, 1 µm, and 500 nm particles. There are two features in Fig. 1-5.

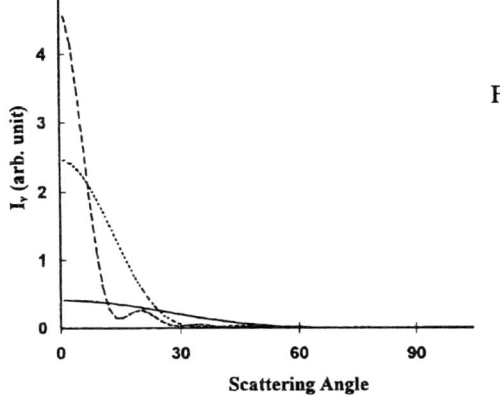

Fig. 1-5: Unit volume angular scattering patterns for 2 µm particles (dashed line), 1 µm particles (dotted line) and 500 nm particles. The particles are polystyrene latex spheres dispersed in water. The light wavelength in vacuo is 633 nm.

The first is that the larger the particles are the larger the absolute scattering intensity per unit volume will be at small scattering angles. As the first order approximation, the scattering per unit volume is proportional to the mass of the particle at small scattering angles. The second feature is that the shape of the angular pattern becomes smoother and less angle-dependent when particles get smaller. Because particles in the sample have different scattering intensity per unit volume at a given scattering angle, the signal detected is weighted by the scattering intensity of the constituents of the sample, not by the volume or weight percentages of these constituents.

In general, a measurement performed at small scattering angles will be heavily weighted by the scattering from the larger particles in the sample and a measurement performed at large scattering angles will be weighted more by the scattering from the smaller particles in the sample. However, because of the minima in the angular pattern, there may be too little scattering from particles of a particular size at the measurement angle. This is especially true for particles of sizes larger than the wavelength of the light in the medium. Fig. 1-6 displays the scattering from a unit volume of particles of different sizes at three scattering angles. For a sample composed of a bimodal distribution of 440 nm and 240 nm particles, there will be little signal from the 440 nm particles if the measurement is performed at 90 degrees.

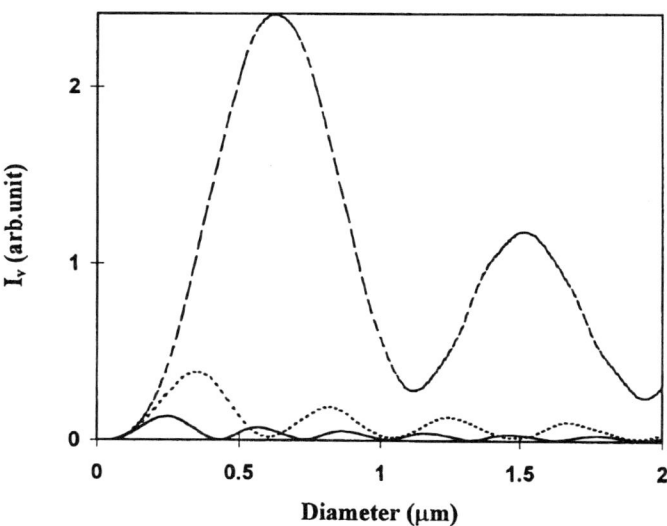

Fig. 1-6: Unit volume scattering from polystyrene latex spheres in water at the scattering angles of 30 degrees (dashed line), 60 degrees (dotted line), and 90 degrees (solid line), respectively.

Obviously, the size distribution obtained will be a distorted one. Thus, although the bias due to different scattering intensities from particles of different sizes may be corrected by changing the weighting of the size distribution retrieved from the ACF from intensity-weighted to volume- or weight-weighted using a proper model (e.g. the Mie theory can be used for solid spheres and various form factors can be used for submicron particles), such correction will not be effective, or will introduce large uncertainties, when information is missed in the ACF due to extremely weak scattering from certain components at the measurement angle. Therefore, if the measurement is only performed at one angle, even if the distribution is reweighted, the result may be still incorrect. Multi-angle measurement is necessary in these cases to eliminate the blind spot effect and to confirm and verify the size distribution. For most samples there is an optimal angular range from which the least biased size distribution can be obtained. For example, from the scattering patterns in Fig. 1-5, it is clear that for a mixture of equal volumes of 500 nm and 2 µm latex particles, the best angular range would be between about 10 to 25 degrees. At scattering angles smaller than 10 degrees, the scattering from 2 µm particles will overshadow the scattering from the 500 nm particles, even using a delay time suitable for 500 nm particles; and vice versa for 2 µm particles at scattering angles larger than 25 degrees. Another application made possible by multi-angle measurement, the opposite of optimizing signals from all constituents as described above, is to choose a specific angle in order not to "see" certain constituents. Examples of such applications are measuring the main component in contaminated samples, detecting trace amount of small particles in samples that are mainly composed of large particles, etc.

The scattering intensity per unit volume is proportional to the mass of the particle in small scattering angles. For solid particles, the mass changes as the cube of the diameter change. For coils of macromolecule chains the mass changes approximately as the square (or less) of the hydrodynamic diameter change. Thus, for the same dynamic size range, the scattering intensity dilemma in angle selection is less pronounced for macromolecule chains than for solid particles.

The second difficulty in PCS measurement of polydisperse samples is that each particle has its own distinctive moving velocity, or rate of diffusion. Decay constants for different particles in the sample may be quite different covering a wide dynamic range. According to Eq.5, the decay constant is inversely proportional to the hydrodynamic diameter of particles, i.e., the dynamic range of the decay constant will be the same as the size range of the sample. Conventional correlators with the delay time arranged linearly will not perform satisfactorily for samples of broad size distribution. With a finite number of delay channels (typically between 256 channels to 1024 channels), the coverage for a wide range of decays just cannot be accomplished. As an empirical rule in choosing a proper delay time in using a correlator with linearly spaced delay time, the product of the mean decay constant $<\Gamma>$ and the maximum delay time τ_{max} should be around 3. If the delay time chosen is too short, slow motion from large particles may not be fully covered. If the delay time chosen is too long, the details of fast motion from small particles will be missed. In either case, error or uncertainty in the size information obtained will be large. For samples having a broad range of decay constants, one either has to make a few measurements using different delay times at the same scattering angle or make a multiangle measurement. To increase the dynamic range, a correlator should have its delay time arranged logarithmically or quasi-logarithmically. Thus, fast motion from small particles can be captured at short delay times where the channels are densely spaced, while slow motion from large particles of the sample can still be observed through those channels at long delay times. For example, in the Coulter N4 Plus multi-angle PCS instrument, 80 correlator channels are divided into eight groups. Channels in each group are spaced linearly while the delay time between the groups are incremented logarithmically. By such arrangement, the 80 real channels span a delay time scale equivalent of 3072 linearly-spaced channels with a better statistical accuracy and a dynamic sizing range of more than 300. Fig. 1-7 is the ACF of the sample in Fig. 1-4 using Coulter N4 Plus. The ACF is plotted in logarithmic scale to show the arrangement of the delay channels.

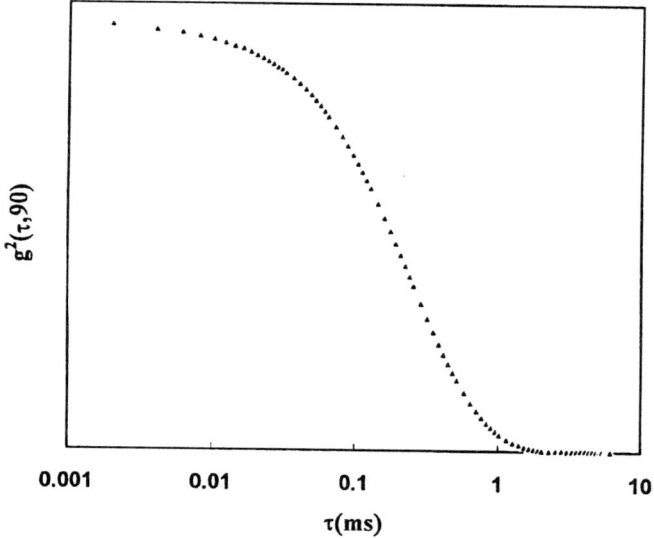

Fig. 1-7: The ACF of the sample in Fig. 1-4 obtained using the Coulter N4 Plus at 90 degrees.

Regardless of the delay channel arrangement, each instrument has its own specified minimum and maximum delay times. Because the characteristic decay is related to the observation angle (see Eq. 1-5), the best coverage for all particles in the sample can be achieved by choosing a proper angle so that the decays will fall into the delay time range of the instrument. For example, if the minimum delay time for a PCS instrument is 0.1 μs and the instrument has 256 linear channels, a measurement performed on 0.5 nm particles at 90 degrees will produce an ACF with a sharp decay in the first few channels while the remaining channels are all near the baseline, which may result in large uncertainty in data analysis. A 60-degree scattering angle measurement in this case will yield an ACF with better overall coverage.

Finally, the third advantage of multi-angle PCS measurement is related to sample concentration effects. In many PCS experiments, the sample used is diluted from the originally concentrated one to meet concentration requirement for PCS measurements: it must be dilute enough that multiple scattering and inter-particle interactions in the Brownian motion can be avoided, and the scattering will not saturate the detection device; yet it must not be so dilute that there is not enough scattering to produce a good signal-to-noise ratio and satisfy the statistics for photon correlation. However, there are many instances where the concentration of the suspension (or solution), or the sample scattering intensity, cannot be adjusted to fit the experimental requirement. Several examples are: a) the concentration is already very low as in many biology samples; b) the original concentration is high but the scattering is still weak as in many surfactants and synthetic polymer samples; and c) the concentration is too high, causing detector saturation, but the sample cannot be diluted otherwise the particles will change. In these examples, one needs to adjust the experimental conditions so that a proper measurement can be made. Changing scattering angle is the most feasible and convenient approach.

The amount of light scattered into the detector is directly proportional to the scattering volume. In most collecting optics, as the one described in Fig. 1-3, the scattering volume at a scattering angle θ is greater than that at 90 degrees by a factor of $\sin^{-1}\theta$. By this relation, for example, about 3.9 times as much light is collected at 15 degrees as compared to at 90 degrees. Since the signal-to-noise ratio is proportional to the square of the light intensity, the signal improves by a factor of 15 over that for 90 degrees scattering angle. For over-diluted or weakly scattering samples, performing the measurement at small angles is equivalent to using a more powerful laser. On the other hand, for samples scattering too much light, measurement at large angles will effectively reduce the scattering intensity reaching the photodetector thus avoiding possible detector saturation.

The time scale of a PCS measurement and its relationship to scattering angle provides another way to understand the advantage of a multi-angle instrument or apparatus. A

PCS measurement can be thought of as a measure of the time it takes for a particle to diffuse a distance K^{-1} (see Eq. 5). This time is proportional to K^{-2}. Thus, the time scale for the diffusion process viewed at 15 degrees is, for example, 29 times as long as that for the same process measured at 90 degrees. In the other words, a low-angle measurement effectively slows down the diffusion process of fast moving particles as viewed from the detector. Consequently, the sample time used for the experiment can be increased by the same factor. Therefore, in this example, 29 times as much light per sample time can be collected if the measurement is performed at 15 degrees instead of at 90 degrees, permitting measurement of weakly scattering samples. Conversely, performing measurements at high angles will effectively speed up the diffusion process enabling measurement of strongly scattering samples by using a small delay time. Thus, by changing the scattering angle one effectively changes the scattering volume and the time scale for the diffusion process, which is equivalent to manipulating the sample concentration without actually changing the sample.

There are two data analysis schemes for multiangle PCS measurement. The simpler one is the analysis of the ACF measured at a single angle, either the cumulants analysis for the mean value and the polydispersity index, or size distribution analysis utilizing one of the Laplace inversion techniques. This type of analysis is useful in finding the best angle at which the least biased result can be obtained. In practice, one of the quick ways to use multi-angle measurement for quality control or trace amount detection is the *Fingerprint* method [9]. In this method, the cumulants analysis is performed on the ACFs of several angles separately; the mean sizes are then plotted versus the scattering angles. Based on the fact that large particles scatter strongly in the forward direction and small particles exhibit a more isotropic pattern, if the sample is polydisperse or multimodal, the mean size value will monotonically decrease as the scattering angle increases. If the sample is from a production line and the measurement is for quality control, then the mean value versus scattering angle pattern, the fingerprint, can be used to monitor the sample quality. Other applications of the fingerprint include colloid stability studies, aggregation or flocculation studies, and adsorption studies. Fig. 1-8 shows fingerprints for mixtures of 100 nm and 1000 nm polystyrene latex particles at different mixing ratios. Each ratio exhibits a characteristic fingerprint shape. In fact, fingerprint shape can be used as an alternate method for sample characterization, in lieu of a size distribution.

There are several experimental error sources that can cause the results obtained at different angles to be different, in addition to the theoretical reasons explained earlier.

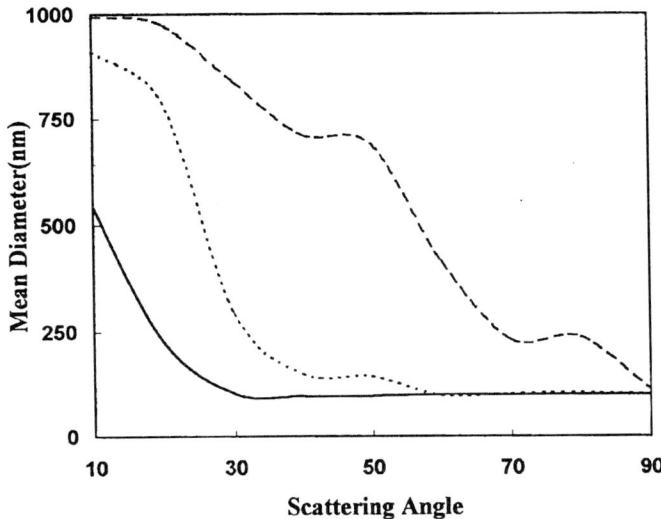

Fig. 1-8: The simulated fingerprints of different mixtures of 100 nm/1000 nm polystyrene particles suspended in water at $\lambda_o=633$ nm. Solid line: 0.1% of 1000 nm; dotted line: 1% of 1000 nm; dashed line: 10% of 1000 nm.

Avoiding and correcting these error sources are vital to obtaining a correct size distribution. Some common error sources are a) incorrect angle values resulting from instrument misalignment, b) a scratch spot on the sample chamber surface producing a strong local oscillator that changes the experiment from the self-beating mode to the homodyne mode, c) incorrect delay time, d) a few foreign particles drifting in and out of the scattering volume causing particle number fluctuation and baseline error, e) a dynamic sample that changes with time.

One big disadvantage of analysis based only on the ACFs from individual angles in a multi-angle measurement is that when the results from different angles are different, even after proper conversions to present the distributions in volume percentage, one is often puzzled as to which is the right answer. In these cases, knowledge of the sample and of the scattering technology is needed to correctly interpret the results. In certain instances, a global analysis that incorporates both the ACFs and the scattering intensities (the term $<I(t,\theta)^2>$ in Eq. 1, measured in a calibrated instrument) from different angles into the inversion process presents a better approach [10-13]

In the global analysis, both the ACFs of several angles and the scattering intensities measured at these angles are used in the fitting process simultaneously. For that purpose, the instrument has to be calibrated for the intensity readings in order to obtain correct intensity values. The matrix format for the ACF from each angle is:

$$\begin{bmatrix} g^{(1)}(\tau_1,\theta_1) \\ ... \\ g^{(1)}(\tau_m,\theta_1) \end{bmatrix} = \begin{bmatrix} e^{-\Gamma_1(\theta_1)\tau_1} & ... & e^{-\Gamma_j(\theta_1)\tau_1} \\ ... & ... & ... \\ e^{-\Gamma_1(\theta_1)\tau_m} & ... & e^{-\Gamma_j(\theta_1)\tau_m} \end{bmatrix} \begin{bmatrix} s(d_1,\theta_1) & 0 & 0 \\ 0 & ... & 0 \\ 0 & 0 & s(d_j,\theta_1) \end{bmatrix} \begin{bmatrix} G(d_1) \\ ... \\ G(d_j) \end{bmatrix} \qquad \text{Eq. 1-9}$$

where G(d) is the volume-weighted size distribution and $S(\theta_i)$ is a j x j diagonal matrix whose diagonal components $s(d_j,\theta_i)$ are the volume-to-intensity conversion factors. The combination of several ACFs will produce several matrices (the number of matrix is equal to the correlator channel number), each of which will have the format:

$$\begin{bmatrix} <I(\theta_1)>^2 g^{(1)}(\tau_m,\theta_1) \\ ... \\ <I(\theta_n)>^2 g^{(1)}(\tau_m,\theta_n) \end{bmatrix} = \begin{bmatrix} K(\theta_1)S(\theta_1) \\ ... \\ K(\theta_n)S(\theta_n) \end{bmatrix} \begin{bmatrix} G(d_1) \\ ... \\ G(d_j) \end{bmatrix} \qquad \text{Eq. 1-10}$$

where $K(\theta_i)$ is a 1 x j matrix whose elements are the kernel of the transform exp(-$\Gamma_j(\theta_n)\tau_m$). A global fitting procedure using all the matrices will yield one size distribution result. Several algorithms have been used for the multi-angle inversion, for example: the multi-angle version of CONTIN [10,12], non-negative least-squares technique [11], and the singular value analysis [13]. Although the algorithms for this type of approach are more complicated than the single angle Laplace inversion, it takes less than one minute to complete an analysis on an IBM compatible 33 MHz 386SX [11]. When compared with the single angle analysis, global analysis using each of the above three algorithms for simulated data and ideal systems such as spherical polystyrene latex has shown improved accuracy (recovering the correct mean values), resolution (resolving individual population from bimodal distribution), and sensitivity (detecting minor components in the sample) for broad distributions. This approach, however, requires prior knowledge of the sample material, such as the refractive index and the shape, in order to choose proper conversion factors (the matrix S). It works well with spherical particles. For other shapes or for the sample with imprecisely known refractive index, good results may not be expected.

In summary, the PCS technology has become a mature tool in particle size characterization for both industrial applications and academic research. Although the size distributions obtained from PCS measurement have low resolution and often distorted shapes due to the nature of information content in the ACF, it is still one of the best non-invasive submicron sizing methods available for measuring particulates in liquid media. Multi-angle measurement using a digital correlator with logarithmically spaced delay times can extend the dynamic size range of the measurement and obtain more correct results by avoiding intensity blind spots in the scattering pattern and optimizing the concentration effect for the measurement. Global analysis combining both dynamic and static light scattering data is a challenge for further advancement of PCS technology. There is still much work to be done before such analysis can be applied to real samples.

1.4 Acknowledgement

The author thanks Coulter Corporation for the support and Mr. Gordon Row of Coulter Corporation for performing the PCS repeatability test and helpful discussion.

1.5 References

[1] *International Standard ISO/DIS 13321*, International Organization for Standardization, Geneva, 1996

[2] R. Pecora (ed.) *Dynamic Light Scattering*, Plenum, New York, 1985

[3] B. Chu, *Laser Light Scattering: basic principles and practices*, 2nd edition, Academic Press, New York, 1991

[4] R. G. W. Brown; R. Jones; J. G. Rarity; K. D. Ridley, *Applied Optics*, 1987, 26, 2 383

[5] D. E. Koppel, *J. Chem. Phys.*, 1972, 57, 4814

[6] S. W. Provencher, *Computer Phys. Comm.*, 1982, 27, 213

[7] B. Chu; R. Xu; S. Nyeo, *Part. Part. Syst. Charact.* 1989, 6, 34

[8] R. Xu, Particle Size Analysis Using Laser Scattering, in *Liquid and Surface Borne Particle Measurement Handbook*, Eds. J. Knapp, T. Barber and A. Lieberman, Marcel Dekker, New York, 1996, Chapter.18, p 745-777.

[9] H. Hildebrand; G. Row, *Am. Lab. News*, 1995, 2, 6

[10] S. E. Bott, In *Particle Size Analysis*, Ed. P. J. Lloyd, Wiley, London, 1988, p 77-88.

[11] G. Bryant; J. C. Thomas, *Langmuir*, 1995, 11, 2480

[12] P. Stepánek, In *Dynamic Light Scattering*, Ed. W. Brown, Oxford University Press, New York, 1993, p 177-240.

[13] R. Finsy; P. D. Groen; L. Deriemaeker; E. Gelade; J. Joosten, *Part. Part. Syst. Charact.*, 1992, 9, 237

Address of the author:
Renliang Xu,
Coulter Corporation
Mail Code: 195-10
P.O. Box 169015
Miami, FL 33116-9015, USA

2 Possibilities and Limitations of Laser Light Scattering Techniques for Particle Size Analysis

Prof. Dr. I. Zimmermann, Würzburg

The electromagnetic theory of light scattering explains how light is scattered by a small particle. If we have the scattering pattern of a small particle this theory allows us to calculate its size. The determination of particle sizes from the scattering pattern of monochromatic light is called light scattering analysis. However the theory of light scattering shall be outlined briefly before discussing this technique, its possibilities and limitations more intensively.

2.1 Scattering by a Single Particle

The following equation describes the propagation of light

$$\vec{E}(r,t) = \vec{E}_0(r) \sin(\omega t - k\vec{r} - \Phi_0)$$ Eq. 2-1

If we define the starting phase by $\Phi_0 = 0$ and consider plane incident waves only then Eq. 2-1 transforms as follows [1-3]:

$$\vec{E}(z,t) = \vec{E}_0(z) \sin(\omega t - k\vec{z})$$ Eq. 2-2

Interactions between light and particles occur if such a plane wave of a wavelength λ (circular frequency $\omega = 2\pi/\lambda$) hits on a small particle of fraction index n and of radius a or diameter d respectively. In essence the electromagnetic field of the light induces in the particle dipoles or multipoles which in return emit electromagnetic radiation, the scattered light (Fig. 2-1).

2.1.1 Mie Theory [3]

An exact solution of the scattering problem was given by Mie at the beginning of this century. He solved the wave equation for a particle of spherical symmetry and obtained the following expression describing the intensity of the scattered light:

$$I = I_0 \frac{F(\Theta, \Phi)}{k^2 r^2}$$ Eq. 2-3

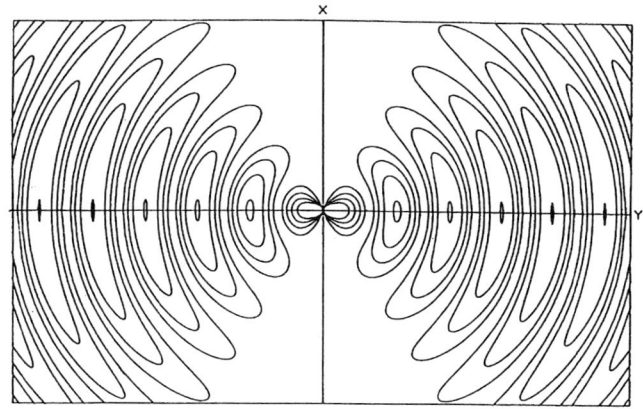

Fig. 2-1: The electromagnetic field of the light induces in the particle dipoles or multipoles which in return emit electromagnetic radiation, the scattered light

The function $F(\Theta, \Phi)$ is dimensionless and independent of direction and distance between scattering particle and screen. However it is strongly dependent on the polarization of the incident light. $F(\Theta, \Phi)$ is a friendly and convenient short hand notation for rather complicated expressions build up by spherical Bessel functions and Legendre polynomials. The evaluation of these expressions results in the following expressions for the intensity I_s of the scattered light:

$$I_s^\parallel = |S_2(\Theta)|^2 \frac{I_0}{k^2 r^2} \qquad \text{Eq. 2-4}$$

and

$$I_s^\perp = |S_1(\Theta)|^2 \frac{I_0}{k^2 r^2} \qquad \text{Eq. 2-5}$$

The expressions $S_1(\Theta)$ and $S_2(\Theta)$ are given by

$$S_1(\Theta) = \sum_{n=1}^{\infty} \frac{2n+1}{n(n+1)} \{b_n \tau_n(\cos\Theta) + a_n \pi_n(\cos\Theta)\} \qquad \text{Eq. 2-6}$$

and

$$S_2(\Theta) = \sum_{n=1}^{\infty} \frac{2n+1}{n(n+1)} \{b_n \pi_n(\cos\Theta) + a_n \tau_n(\cos\Theta)\} \qquad \text{Eq. 2-7}$$

a_n and b_n are so called multipole coefficients. Their explicit calculation is extremely time consuming. In essence they are determined by the wavelength of the incident light, the relative fraction index m between the spherical particle of diameter d and its surrounding.

π_n and τ_n are defined by the following recursion:

$$\pi_n = \frac{2n-1}{n-1}\cos\Theta\,\pi_{n-1} - \frac{n}{n-1}\pi_{n-2}$$

$$\tau_n = n\cos\Theta\,\pi_n - (n+1)\pi_{n-1} \qquad \text{Eq. 2-8}$$

The starting values of the recursion are: $\pi_0 = 0$ and $\pi_1 = 1$.
The scattering angle Θ measures the deflection of the beam scattered by a particle and the direction of the incident light. The intensity of the scattered light is strongly dependent on this scattering angle.

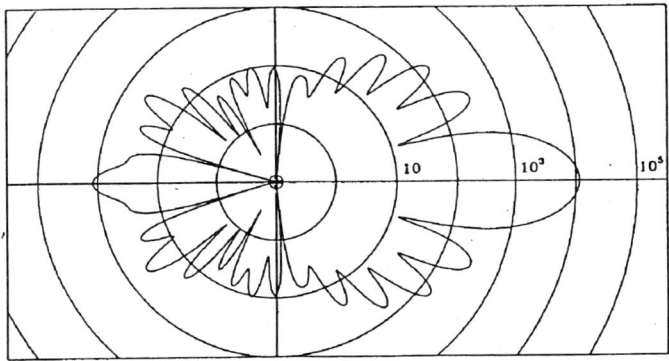

Fig. 2-2: In the graph the concentric circles represent points of equal scattering intensity. The distribution of light scattered by a small water droplet of a diameter of 3 µm was calculated using a program worked out by A. Naqwi [from 4].

2.1.2 Mie approach $\frac{\lambda}{10} \leq d \approx \lambda$:

According to the Mie - theory microparticles are able to scatter light if the ratio of the wavelength of the incident light and of the circumference of the particle is exactly defined in a way allowing for the formation of standing waves inside the particle. For particles with $\frac{\lambda}{10} \leq d \approx \lambda$ the scattering of light has to be calculated using the Mie approach.

2.1.3 Rayleigh approximation $d \leq \lambda$:

For particles with $d \leq \lambda$ the scattering properties can be described by a much simpler approximation.
For particles with $d \leq \lambda$ the phase of the incident light is the same over the whole particle. This means in contrast to the range of particle sizes described by the Mie

approximation there are no interferences of light within the particles. If we are using light of a wavelength of $\lambda = 750$ nm this condition is fulfilled for particle sizes up to 70 nm. The intensity of the scattered light is given by:

$$I = I_0 \frac{\lambda^2}{4\pi^2}\left(\frac{\pi d}{\lambda}\right)^6 \left|\frac{m^2-1}{m^2+2}\right|^2 (1+\cos\Theta) \qquad \text{Eq. 2-9}$$

2.1.4 Fraunhofer approximation d >> λ

If the wavelength of the light incident into a sample is much shorter than the diameter of the particles the Fraunhofer approximation describing light scattering by means of a geometrical optics approach can be used. The intensity of the scattered light depending on the scattering angle Θ, the particle size d and the wavelength λ is given by:

$$I = I_0 \frac{\sin^2\left(\frac{\pi d}{\lambda}\sin\Theta\right)}{\left(\frac{\pi d}{\lambda}\sin\Theta\right)^2} \qquad \text{Eq. 2-10}$$

In all approximation procedures the intensity of scattered light is proportional to

$$\alpha = \frac{\pi d}{\lambda} \qquad \text{Eq. 2-11}$$

This parameter is called „Mie - factor".

2.2 Scattering by Collective Samples

If the incident light hits on a collective sample of dispersed particles each particle scatters light independent from all others. The intensity of the scattered light at a given position on the detector is given by the sum of the contributions of all scattering particles.

2.2.1 Measurement of scattered light

In these days lasers are used as powerful monochromatic light sources in scattering experiments. The light scattered by collective samples of particles is focused on detectors by means of so called Fourier lenses. In the forward range of scattering, this means at small scattering angles, this results in an intensity distribution concentric to primary beam of incident light. The intensity of the scattered light is determined by the number of scattering particles in the sample. The position of minima and maxima of the intensity of scattered light are dependent on the size of the scattering particles. Thus the intensity distribution is characteristic for the distribution of particle sizes.

2.2.2 Determination of particle sizes from scattering experiments

The intensity $I(\Theta)$ of scattered light measured by a detector under an angle Θ comprises the contributions of all particles in the cell which interacted with the incident light. Therefore

$$I(\Theta) = \int_0^\infty I(a,\Theta) p(a) \, da \qquad \text{Eq. 2-12}$$

$I(a,\Theta)$ is the intensity of the light emitted under an angle Θ by a particle of radius a. It is calculated using one of the approximations discussed above. $p(a)$ is the density distribution of the particles in the sample. The intensity of the scattered light is measured by an finite number of detectors. Each of them measures the light of a given range of deflection angles corresponding to a well defined range of particle sizes. The integral equation, Eq. 2-12, is then approximated by a system of linear equations. Using matrix notation this can be written as follows:

$$\begin{bmatrix} f_1 \\ . \\ . \\ . \\ f_n \end{bmatrix} = \begin{bmatrix} c_{11} & . & . & . & c_{1n} \\ . & . & . & . & . \\ . & . & . & . & . \\ . & . & . & . & . \\ c_{n1} & . & . & . & c_{nn} \end{bmatrix} \begin{bmatrix} x_1 \\ . \\ . \\ . \\ x_n \end{bmatrix} \qquad \text{Eq. 2-13}$$

The column f represents the vector build up by the fluxes of light as measured by the various detectors and weighted in an appropriate way. The coefficients of the c - matrix are calculated as solutions of equation 12 for specific ranges of scattering angle Θ and well defined fractions of particle size. The elements of the column x represent the particle size distribution. They are to be determined. This is done applying the rules of matrix calculation, this means the system of linear equations given by Eq. 2-13 is multiplied from the left by the inverse matrix of c. As mentioned already to calculate the elements of the c - matrix according to equation 11 one of the expressions Eq. 2-3, Eq. 2-9 or Eq. 2-10 is used.

2.2.3 Problems in the particle size analysis using scattered light

Light is emitted by small scattering particles under large scattering angles Θ, the larger the particles the smaller the scattering angles. The intensity of scattered light decreases extremely with increasing scattering angles. This holds especially for particles being smaller than the wavelength of the light used for the scattering experiment.

This problem can be overcome by using the so called PIDS technique. PIDS stands for Polarization Intensity Differential Scattering. As shown when discussing the Mie approximation, the intensity of scattered light is strongly dependent on the polarization of the incident light. Interestingly this dependence on the polarization of the incident light is more pronounced with smaller particles than with the bigger ones. Exploiting this effect allows for a much better resolution of the signals emitted by small particles. If the incident light is consecutively polarized in a vertical and a horizontal direction the PIDS signal is determined by the difference of the corresponding intensities of scattered light.

$$PIDS = I_{s,v} - I_{s,h} \qquad \text{Eq. 2-14}$$

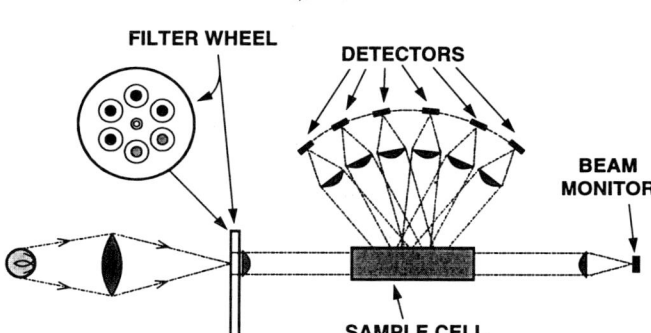

Fig. 2-3: PIDS system

As can be seen from Fig. 2-4 in the range of scattering angles from 70 to 110 degrees there are significant differences in the PIDS signal produced by the various particle sizes. To improve the reliability of the results of the measurements they are performed using three different wavelength (e. g. 450, 600 and 900 nm).
Technically this is achieved by filtering out light of three different wavelengths from the spectrum of an incandescent lamp being vertically and horizontally polarized. The intensity of the scattered light is measured under the angles of 70, 80 , .., 110 and 140°. The evaluation of these signals is performed using the approximations discussed above. This results again in a system of linear equations which than is added to the already existing equations. Using the matrix notation we obtain the following expanded system of equations [5]:

$$\begin{bmatrix} f_1 \\ \cdot \\ \cdot \\ \cdot \\ f_m \\ p_1 \\ \cdot \\ p_k \end{bmatrix} = \begin{bmatrix} c_{11} & \cdots & c_{1n} \\ \cdot & & \cdot \\ \cdot & & \cdot \\ \cdot & & \cdot \\ c_{m1} & \cdots & c_{mn} \\ b_{11} & \cdots & b_{1n} \\ \cdot & & \cdot \\ b_{k1} & \cdots & b_{kn} \end{bmatrix} \begin{bmatrix} x_1 \\ \cdot \\ \cdot \\ \cdot \\ x_n \end{bmatrix} \qquad \text{Eq. 2-15}$$

It is solved according to the rules of matrix calculation.

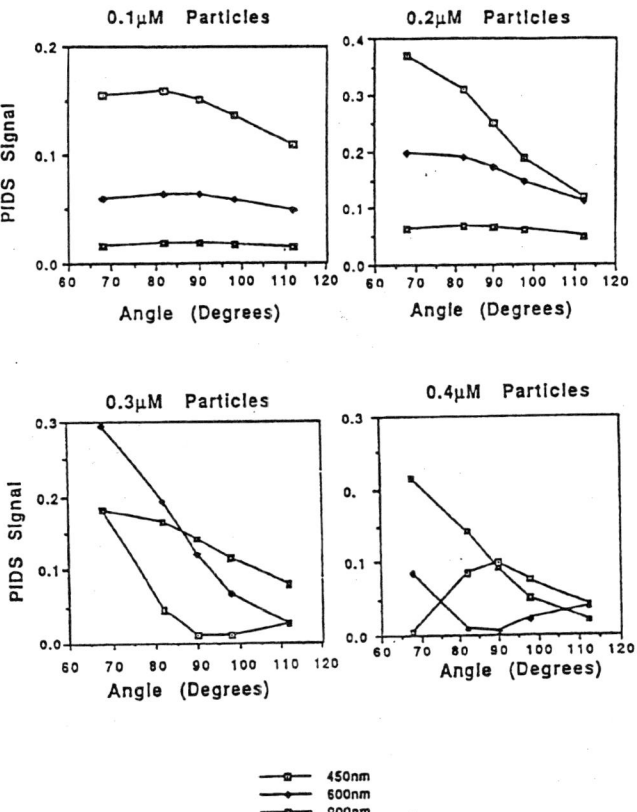

Fig. 2-4: PIDS signals obtained from various particle sizes with different wavelengths.

As the overall dynamics of intensity of scattered light is less pronounced in the PIDS range compared to the conventional scattering experiments. Therefore by separate determination of the intensity of light scattered by the fines of a sample much more sensitive detectors can be used. This allows for a significant expansion of small particle sizes to be determined by using light scattering techniques. This principle is implemented in the Coulter particle sizers of the LS series.

Another problem in light scattering experiments are the dimensions of the detector arrays. The radian of the light emitted by a large particle is very small. Therefore the signals have to be determined with a very high radial resolution. To be able to determine the size of particles of 100 μm precisely one must be able to measure values of the angles Θ of 0.02 ° in increments of 0.002 °. With detectors having dimensions in the μ range one has reached the upper limit of the method using conventional Fourier lenses to focus the scattered light on the detector. Increasing the focal length of the

Fourier lens offers an approach to allow for a reliable determination of larger particle sizes. This however results in a correspondingly longer axis of the optical bench of the system combined with a series of new difficulties like calibration problems due to vibrations or eventually the size of the instrument.

The use of „Reverse Fourier" lenses may offer a way out of this dilemma. These lenses allow for an expansion of the overall focal length of the lens without the need for a bigger instrument. Using this kind of lenses allows for the determination of angles of 0.01° in increments of 0.001°. Thus the range of particle size can be expanded up to 2000 µm.

2.3 Discussion

Progress in the field of data processing as well as in the development of highly efficient and extremely small sized detectors had a tremendous impact on the possibilities of particle size analysis by laser light scattering methods. Equipment presently commercially available at a reasonable price offers the possibility to determine particle size distributions in the range from 0.04 - 2000 µm. This is achieved by simultaneous application of PIDS technique.

2.4 References

1. M. Kerker: The scattering by small particles; Academic Press, Orlando 1987

2. H. C. van de Hulst: Light scattering by small particles; Dover Publications, N. Y. 1981

3. J. Popp: Elastische und inelastische Lichtstreuung an einzelnen spärischen Mikropartikeln; Dissertation Universität Würzburg 1994

4. M. Stieß: Mechanische Verfahrenstechnik Bd.1, p. 109; Springer Verlag, Heidelberg 1992

5. R. L. Xu, Coulter Corp. Miami,FL: persönl. Mitteilung

Address of the author:
Prof. Dr. Ingfried Zimmermann
Institut für Pharmazie und Lebensmittelchemie
Am Hubland
D-97074 Würzburg

3 Improvements in Particle Size Analysis Using Light Scattering

Renliang Xu, Coulter Corporation, Miami

3.1 Introduction

With global demands for quality control and quality assurance in both industrial processes and research and development, and with more and more industries adopting the ISO 9000 standard series, the characterization of particulate materials is becoming more and more important. The results of characterization determine and influence other portions of the life cycle of particulate materials, such as manufacturing, storage, transportation and ultimate application. In many properties of a particulate material, the mean size and the size distribution are often the most important parameters that determine other macroscopic properties of the material and thus characterize the material. The uniqueness of the huge numbers of particles per unit volume or per unit weight (for example, one cubic centimeter contains only one cubical particle 1 cm long but one million particles 100 μm long, and one trillion particles 1 μm long) and the polydisperse nature of many industrial particulate systems make the methods for sizing small particles with dimensions of a millimeter or smaller completely different from the methods used for sizing macroscopically chunky or bulky materials.

There are ensemble methods and non-ensemble methods used in particle sizing depending on whether each measurement detects or senses the size from an ensemble of particles in the sample. For non-ensemble methods, materials have to be separated or fractionated into separate components according to a certain property of the material prior to the measurement. All non-ensemble sizing methods are composed of a separation or fractionation mechanism. Two typical methods used to size one particle at a time are the optical particle counter and the Coulter counter. Typical ensemble methods for particle size determination are photon correlation spectroscopy and elastic light scattering. The advantages of these ensemble methods are broad dynamic range and fast and non-intrusive measurement. The disadvantages are their low resolution and model-dependent data retrieving processes when compared with a single particle counter. The choice of a proper analytical method for particle sizing usually depends on the requirement of the application and accessibility to a suitable analytical technique.

In the past two decades, light scattering has become *the* most popular and important physical chemical mean for particle sizing. There are several branches in the physical sciences that use light scattering phenomena to study various properties of materials. Tab. 3-1 lists the common scattering technologies used in material study. Each of these technologies has its own terminology, expertise, applications, and techniques of measurement, and involves different disciplines in the physical sciences. Several of the above scattering technologies can be used to measure particle size (the non-italic ones in Tab. 3-1). The elastic light scattering intensity measurement based on the time-averaged angular intensity pattern scattered from dispersed materials (the one with the bold font in Tab. 3-1) has gained wide acceptance in the past decade for sizing a broad range of air-borne and liquid-borne particulate materials due to advancements in laser and computer technology and the subsequent development of commercially viable instruments. In the following, we focus our discussion only on angular intensity measurement without further mentioning other scattering technologies.

Tab. 3-1: Common Light Scattering Technologies

INTENSITY FLUCTUATION	Photon Migration Measurement
	Transient Scattering
	Photon Correlation Spectroscopy
	Diffusive Wave Scattering
AVERAGED INTENSITY	**Angular Scattering**
	Turbidity Measurement
FREQUENCY ANALYSIS	*Phase Doppler Analysis*
	Raman Scattering
	Brillouin Scattering
UNDER AN APPLIED FIELD	*Electric Field Scattering*
	Electrophoretic Scattering
	Scattering Under a Flow Field

In light scattering measurements one obtains the mean particle size and particle size distribution through a matrix conversion of scattered intensity measurements as a function of scattering angle and wavelength of light based on applicable scattering theory. This is an absolute method. Once the experimental setup or instrumentation is correct, calibration or scaling is not necessary in order to obtain the volume (or weight) percentage of each component. The measurement is fast and non-intrusive and can cover an extremely broad size range: spanning almost five orders of magnitude from tens of nanometers to a few millimeters.

Due to the immaturity of the technology and lack of computing power, in the early days of light scattering sizing, only the Fraunhofer diffraction theory was used for calculating particle sizes of a few micrometers and larger. For that reason, the angular light scattering intensity measurement is historically known throughout industry as *laser diffraction*. This term has the advantage in that it distinguishes itself from al other scattering technologies, and has also been accepted by national and international particle characterization communities. However, *laser diffraction* no longer reflects the current state of the technology because of the advancement of this technology during the past few years:

1) light scattering is not limited only to simple diffraction effects. More general approaches based on the Mie scattering theory from measurement of scattering intensity in a wide scattering angular range are employed. The size range has been extended into the submicron region; and

2) a light source with continuous wavelength (a white light) is used to complement the main laser source in order to gain more characteristic information for submicron-size particles based on the wavelength and polarization dependence of the scattering intensity. Thus, *light scattering* may be a more proper term for this technology.

In this chapter, we will discuss in details the wavelength and polarization effects in the improvements of particle sizing using scattering measurement for submicron-size particles. To maintain the integrity of the chapter, the principles of angular intensity light scattering from a single particle will be introduced in Sec. 3.2. The principles and the advantages in using the wavelength and polarization effects in particle sizing are the content of Sec. 3.3. A typical instrumental design incorporating measurement of scattering intensity at different angles, wavelengths and polarizations and several examples to demonstrate the enhancement in particle sizing from such a design are included in Sec. 3.4. A brief summary is given in Sec. 3.5. The content of this chapter will be such chosen that a novice with a college background will be able to understand the technology as it is used as a tool in particle characterization. To that end, all derivations of the related equations, as well as many technical details, have been omitted so that readers will learn the essence of the technology without getting lost into the maze of mathematics. For advanced readers, this chapter will serve as a useful reference from which the basic formulas of light scattering theory, the present state of the art of light scattering instrumentation, and in which the sources for further reading may be found.

3.2 Angular Light Scattering Intensity

Light is electromagnetic radiation in the frequency range from approximately 10^{12} Hz (infrared) to 10^{17} Hz (ultraviolet). When light illuminates a material, e.g., a particle, which has a dielectric constant different from unity, depending on the wavelength of the light and the optical properties of the particle, the light will be absorbed, or scattered, or both. The absorbed light energy that becomes the excitation energy of the particle will be dissipated mostly through thermal degradation, i.e., converted to heat, or lost through a radiative decay producing fluorescence or phosphorescence depending on the electronic structure of the particle. When light interacts with the electrons bound in the material that re-radiate light, scattering is observed. Because most materials exhibit strong absorption in the infrared and ultraviolet regions which greatly reduces scattering intensity and thus is to be avoided, most light scattering measurements are performed using visible light of wavelengths from 350 nm to 900 nm. The scattering intensity from a unit volume that is illuminated by a unit flux of light is a function of the complex refractive index ratio between the material and its surrounding medium. When the particle is much smaller than the wavelength of light, its scattering intensity is inversely proportional to the fourth order of the light wavelength, i.e., the shorter the wavelength, the stronger the scattering. This is the reason that why the sky is blue at midday and red at sunrise or sunset because one sees the scattered sunlight during daytime and sees the transmitted sunlight during dawn and dusk. The gorgeous blue color of ocean can also be attributed in large extent to the strong scattering of the short wavelength portion in the sunlight. The deeper the water level, the deeper the blue color becomes because of the reduced scattering from the longer wavelength portion of the sunlight. The scattering principle is even used in the traffic control where people use red as the color for the stop light and warning signs because red light has the least scattering power in the visible light spectrum, thus allowing the transmitted light to go through fog, rain, and dust particles and reach the intended detector: in this case the human eye.

For particles that are much bigger than the light wavelength, from synthetic or natural macromolecules, such as proteins, gels, DNAs, and latexes, to minute pieces of bulk materials, such as metal oxides, sugar, pharmaceutical powders, paint, or even the non-dairy creamer one puts in coffee, the scattered light intensity is a function of the following variables: particle dimension, particle refractive index, medium refractive index, light wavelength, direction of polarization, and scattering angle. The scattered intensity from a particulate sample is, in addition to the above variables, a function of particle concentration and particle-particle interaction. In characterizing particle size using light scattering, one optimizes sample concentration to a proper range so that the sample will scatter enough intensity to enable the measurement to be completed with a desired signal to noise ratio, but have minimal particle-particle interaction and minimal multiple scattering (i.e., when the light strikes one particle and is then deflected again by another particle), and not scatter so much as to saturate the detecting system. Thus, in a measurement where the light wavelength and polarization are chosen and the sample refractive index is known, the scattered intensity is only a function of scattering angle, particle shape, and particle size, provided that the refractive index and density of particles in the sample are uniform, which is true for most particulate systems. If the

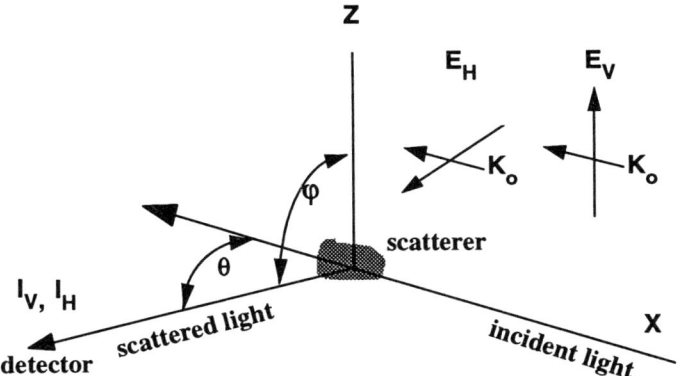

Fig. 3-1: A schematic of light scattering geometry.

relations between scattering intensity, scattering angle, particle shape and particle size are known, one is able to a resolve size distribution for particles of a particular shape from the measured angular scattering intensity pattern.

Fig. 3-1 is a typical scheme for light scattering geometry in the laboratory frame. In the figure, K_0 is the vector of the incident light of wavelength λ propagating along the direction X with the electric vector oriented either perpendicular to the hypothetical table top in the Z direction (termed E_V) or parallel to the table top (termed E_H). The light is scattered in a direction at an angle θ with respect to X and an angle φ with respect to Z. In most light scattering setups, the angle φ is set to 90° for theoretical simplification and practical convenience, and the angle θ is called the scattering angle. I_V and I_H are the scattering intensities detected with the electric vector of the light beam oriented in E_V direction and E_H direction, respectively. In the above arrangement, the magnitude of momentum-transfer vector K ($=|K|=|K_0-K_s|$) is:

$$K = \frac{4\pi}{\lambda} \sin(\frac{\theta}{2}) \qquad \text{Eq. 3-1}$$

The scattering intensity detected at a far field, i.e., when the distance between the detector and the scatterer is much larger than the dimension of the scatterer, is the result of the scattered electric field from different portions of the scatterer. Because scattered fields from different portions of the scatterer interfere with each other, one has to solve the complete set of Maxwell's equations for the entire scatterer in order to obtain the scattering intensity at any given scattering angle. Rigorous analytical formulas for solving Maxwell's equations for a scatterer exist only for a sphere in the spherical coordinate (known as Mie theory after Gustav Mie [1]) and an oriented rod of infinite length in the cylindrical coordinate. There are approximate solutions for some regular shaped scatterers such as ellipsoids and lately for irregular shaped particles[2,3] with very few applications that mostly are limited in academic research. For particles with dimensions either smaller or much larger than the incident light wavelength, further approximations can be applied for those having regular shapes: for small ones, instead of rigorous solutions scattering form factors for particles of a particular shape can be used to describe the interference effect of the scattered light intensity; for large ones, different diffraction formulas are available for non-transparent regular shaped particles. Fraunhofer diffraction is the best known diffraction formula for an opaque disk that has the same diffraction effect as a sphere having the same diameter. Because in a light scattering measurement, all particles are continuously circulated into the measuring chamber, their tumbling and rotational motions make them resembling spheres even though they may be actually irregular in shape, except when non-sphericity is very severe such as long rods or thin disks. In the following, we will only discuss the scattering theories for spheres that are practically useful and have been used in particle characterization.

3.2.1 Mie Theory

The scattering angular patterns of spheres are characterized by scattering minima and maxima at different locations depending on the properties of the particle. Fig. 2 shows the first few scattering lobes for two spherical particles of different sizes.

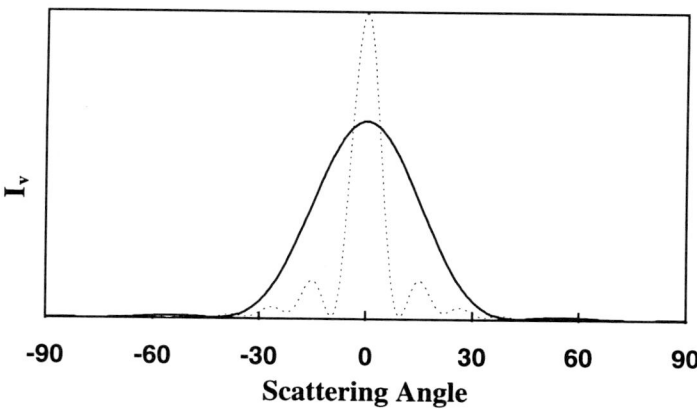

Fig. 3-2: Angular scattering intensity pattern for spheres.

As shown in Fig. 3-2, a general characteristic of scattering from spheres is that the location of the first intensity minimum is closer to the axis and the peak intensity is higher for a large particle (the dot line in Fig. 3-2) when compared with a smaller particle (the solid line in Fig. 3-2). These patterns are axially symmetric with respect to the axis of the incident light; i.e., the scattering pattern is the same for the same absolute value of the scattering angle at the same scattering plane. Thus, the intensity curve at the left side (the negative scattering angle) is a mirror image of the right side (the positive scattering angle) in Fig. 3-2. We need to focus the discussion only on one side of the incident light using all positive angles for convenience.

Mie theory is the rigorous solution for the scattering from a sphere. The derivation of Mie theory can be found in several references [4]. The final analytical formulas of Mie theory are in the form of the spherical Bessel function, which has an oscillating nature:

$$I_V \propto \left| \sum_{j=1}^{\infty} \frac{2j+1}{j^2+j} (a_j \pi_j(\cos\theta) + b_j \tau_j(\cos\theta)) \right|^2 \quad (2) \qquad \text{Eq. 3-2}$$

$$I_H \propto \left| \sum_{j=1}^{\infty} \frac{2j+1}{j^2+j} (a_j \tau_j(\cos\theta) + b_j \pi_j(\cos\theta)) \right|^2 \quad (3) \qquad \text{Eq. 3-3}$$

where a_j and b_j are functions of $\alpha(=\pi d n_0/\lambda_0)$, $\beta(=\pi d m/\lambda_0)$ and τ_j and π_j are functions of $\cos\theta$.

Fig. 3-3 is a 3-D display of the I_V scattering intensity of a unit volume from spheres. The intensity is plotted as a function of scattering angle and sphere diameter. Please note that all three axes are logarithmically scaled in order to cover a wide dynamic range and to reveal details of the intensity variation. There are two general features in the scattering pattern. The first one is the trend of the scattering intensity variation: the intensity is higher for larger particles at the same scattering angle in small angle region and the intensity gets lower as the scattering angle becomes higher for a given size of particle. The trend for angular dependence is reduced when the sphere size gets smaller. For spheres in the nanometer range the intensity becomes much less angular dependent until it reaches a flat line for spheres smaller than 50 nm. The second feature is that due to the oscillatory nature of the Bessel function there are sharp intensity changes from the peak values that center around zero degrees to the first minimum and to sequential minima and maxima, systematic ripples, peaks and valleys. The pattern is characteristic for a given size particle. As shown in Fig. 3-3, if both scattering angle and diameter are spaced logarithmically, the first minimum displays a linear decrease in angle as particles get larger. It is from these features in the scattering pattern that the size information for spherical particles can be resolved through light scattering measurements. The scattering from a different polarization, i.e., the I_H scattering, has different pattern as shown in Fig. 3-4. The most distinguished feature in

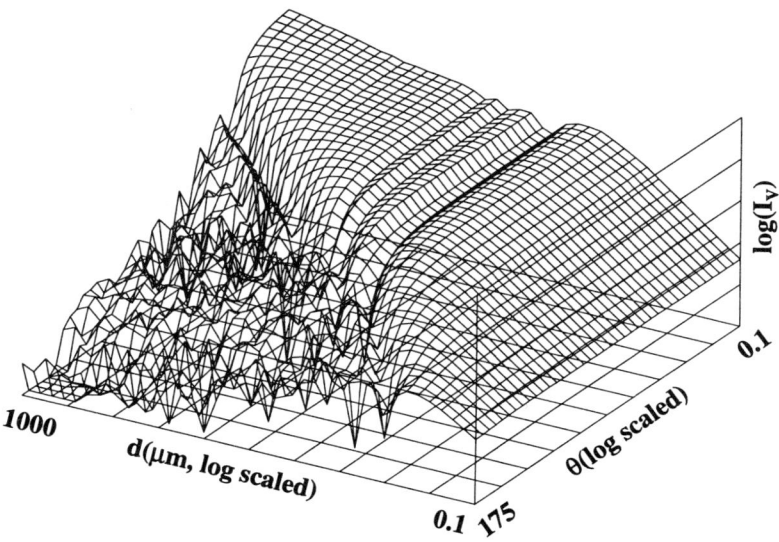

Fig. 3-3: A 3-D display of the Mie scattering intensity I_V from a unit volume of spheres with the relative refractive index m = 1.50 +0i. The light wavelength is 750 nm.

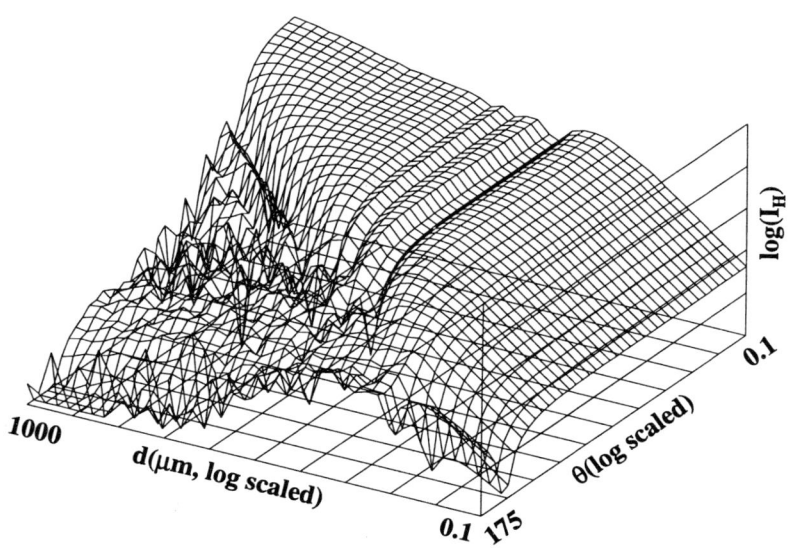

Fig. 3-4: A 3-D display of the Mie scattering intensity I_H from a unit volume of spheres with the relative refractive index m = 1.50 +0i. The light wavelength is 750 nm.

Fig. 3-4 is that there is a dip at around 90 degrees for sub-micron size particles. We will further discuss the differences between I_V and I_H scattering and how to utilize these differences to enhance the particle sizing results in Sec.III. The shape of the scattering pattern varies depending on the values of the relative refractive index m and the incident light wavelength, but these two general features always exist. Fig. 3-5 shows several scattering intensity curves for a 0.5 μm PSL sphere suspended in water and illuminated by light of different wavelengths at different polarizations.

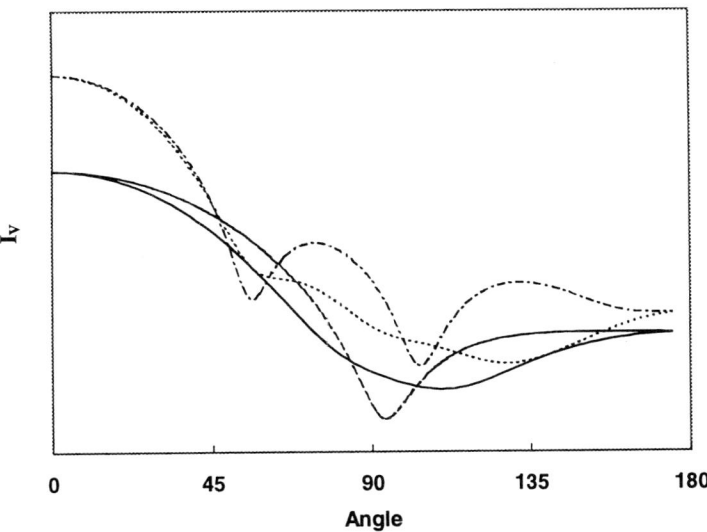

Fig. 3-5: Scattering intensity pattern for a 0.5 mm PSL sphere in water illuminated by different light sources. Solid line: I_H from a diode laser (λ_o = 750 nm); dashed line: I_V from the diode laser; dotted line: I_H from an Ar ion laser (λ_o=488 nm); dot dashed line: I_V from the Ar ion laser.

For large particles, scattering intensities are concentrated at small angles, typically smaller than 10 degrees. Their scattering is mainly from the diffraction effect of the particle edges. Thus, the scattering pattern is centrally symmetric instead of axially symmetric, i.e., it displays concentric rings in the direction of the incident light. In this range, the intensity will be the same for the same solid angle θ with respect to the incident light, and is not restricted to the angle θ in the scattering plane defined by the table top as shown in Fig. 1. The intensity is also insensitive to the direction of light polarization. For scattering in this region, the Eq. 3-2 and Eq. 3-3 can be greatly simplified leading to the Fraunhofer formulation.

3.2.2 Fraunhofer Theory

For particles whose dimension is much larger than the wavelength of light or the materials are highly absorptive, the edge effect of particles contributes more and more to the total scattered intensity. The interference effect now comes mainly from bending of the light at the particle boundary: the diffraction effect. Diffraction can be described by Huygen's Principle: each point on a propagating wavefront is an emitter of secondary wavelets. The locus of expanding wavelets forms the propagating wave. Interference between the secondary wavelets gives rise to a fringe pattern which rapidly decreases in intensity with increasing angle from the initial direction of light propagation. There are Fresnel diffraction for a point source and Fraunhofer diffraction for a parallel beam. In a light scattering measurement, because the light source is always far away from scatterers (compared with the light wavelength) and the optics are usually designed such that the incident beam illuminating the scatterers is homogeneously parallel, only Fraunhofer diffraction will take place. For spheres, the Fraunhofer diffraction is Mie theory at the limiting condition of $d \gg \lambda$:

$$I \propto 2 \left(\frac{\pi d J_1 (\pi d \sin \theta / \lambda)}{\lambda \sin \theta} \right)^2 \qquad \text{Eq. 3-4}$$

where J_1 is the Bessel function of the first kind of order unity.

In particle sizing, Fraunhofer theory can be used for particles that are

1) much larger than the wavelength of the light (typically > 30 µm) and non-transparent, i.e., the particles have different refractive index values than that of the medium (typically with the relative refractive index value m being larger than 1.2), or,

2) highly absorptive (typically with absorption coefficients higher than 0.5).

Eq. 3-4 is derived assuming that a particle scatters as if it were an opaque circular disc of the same projected area positioned normally to the axis of the beam. Because of this assumption, in Fraunhofer theory the refractive index of the material is irrelevant. Eq. 3-4 is valid only for spherical particles. Although various diffraction equations for other regularly shaped particles are available, these formulas have not been found with practical usefulness in particle sizing [5,6]. Because for large particles where Fraunhofer theory applies, the scattering intensity is concentrated in the forward direction, Fraunhofer diffraction is also known as forward scattering. In the Fraunhofer diffraction the angle for the first minimum of scattering intensity is simply related to the size by:

$$\sin \theta_{1st \; minimum} = \frac{1.22 \lambda}{d} \qquad \text{Eq. 3-5}$$

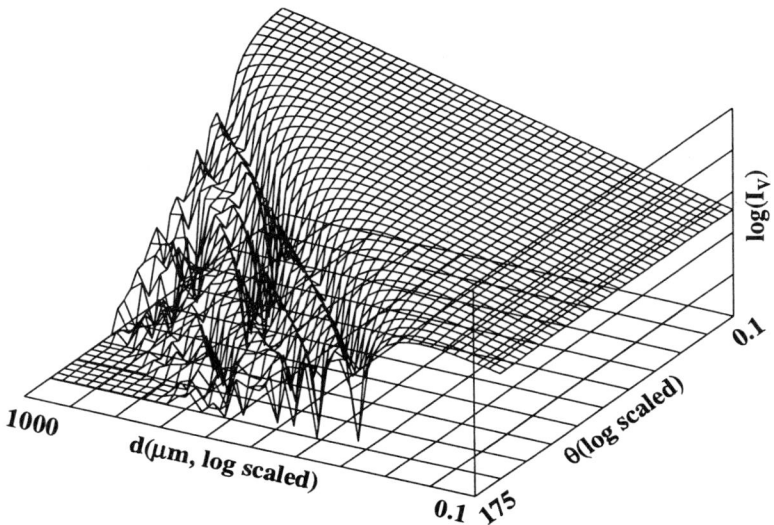

Fig. 3-6: A 3-D display of the Fraunhofer scattering intensity pattern from an unit volume of spheres. The light wavelength is 750 nm.

and most of scattering intensity is concentrated in a very sharp central lobe, which provides a much simpler solution applicable to sizing large particles in a light scattering measurement.

Fig. 3-6 is a similar 3-D display as Fig. 3-1, the unit volume scattering intensity as a function of scattering angle and sphere diameter but by using Eq. 3-4 instead of Eq. 3-2. For the regions of large particles and small angles these two figures are identical which demonstrates that Mie theory can be approximated by Fraunhofer theory.

For particles with smaller m values, even though they may be very big, Eq. 3-4 will no longer be enough because the effects of light transmitting, or refracting through the particles also has to be taken into account. In this anomalous diffraction regime either Mie theory has to be used or other terms accounting for refraction and transmitting effects have to be added to Eq. 3-4 [7].

In the above discussion, the assumption is that the incident light is a parallel beam with a plane wavefront and of homogeneous intensity. If the beam intensity is Gaussian or the light is divergent integral forms of the above formulas that integrate all beam intensity variations have to be used which will make data process and size retrieving much more difficult. For that reason, one of the important targets in designing the optics for a light scattering instrument is to make the beam intensity homogeneous and to reduce the beam divergence to a minimum. In the above equations and plots only the scattering intensity from a unit volume is computed and displayed. The unit volume scattering pattern is the foundation for particle size distribution analysis. Because particles with a certain size have their own characteristic scattering pattern the absolute intensity that they contribute to the measured intensity is proportional to the total volume of these particles. By decomposing the measured intensity pattern into individual patterns corresponding to different sizes the relative amplitudes of these sizes, the size distribution, can be obtained.

Most industrial particles closely resemble spheres and the scattering effect from corners and edges of these particles are averaged out due to tumbling and rotational motion in sample circulation during the measurement. Thus, we can apply Mie theory or Fraunhofer theory to most practical systems with one parameter: diameter, provided that the relative refractive index, the light wavelength and polarization are known. However, such treatment only yields apparent values. One should always keep it in mind that the "size" obtained from most particle sizing technologies, with no exception for light scattering measurement, may be different from the real dimension. The spherical modeling approach is the only feasible choice for a commercial instrument that will be used for a broad range of samples having any shapes.

Once one obtains the scattering angular pattern, a least-square fitting to the formulas of either Mie theory or Fraunhofer theory will yield the sphere diameter for monodisperse particles. For polydisperse samples each individual particle will contribute its scattering to the angular pattern differently. The intensity detected by a detector located at a scattering angle, will be an integration of the scattering from all particles. The scattering pattern then can be represented by the following matrix:

$$\begin{bmatrix} f(\theta_1) \\ ... \\ ... \\ f(\theta_m) \end{bmatrix} = \begin{bmatrix} a(\theta_1,d_1) & ... & ... & a(\theta_1,d_n) \\ ... & ... & ... & ... \\ ... & ... & ... & ... \\ a(\theta_m,d_1) & ... & ... & a(\theta_m,d_n) \end{bmatrix} \begin{bmatrix} x(d_1) \\ ... \\ x(d_n) \end{bmatrix} \quad \text{Eq. 3-6}$$

The left side of the matrix is the detected intensity flux at scattering angle θ_i, i from 1 to m. The flux is from the scattering of $x(d_j)$ percent of particles with diameter d_j that scatter $a(\theta_i,d_j)$ amount of light per unit volume, j from 1 to n. The matrix $a(\theta_i,d_j)$ is the scattering kernel which can be computed from either Mie theory or Fraunhofer theory described earlier. From the measured $f(\theta_i)$ values and the computed $a(\theta_i,d_j)$ values, and by using a matrix conversion technique combined with a fitting algorithm, $x(d_i)$, the particle size distribution, can be resolved. Every instrument has a specified size range that is classified into a certain number of size channels with assigned channel locations and widths, and every instrument has a unique detector array with fixed angular locations and angular coverage for each detector. Thus, a database of $a(\theta_i,d_j)$ with different refractive index values can be built for a particular instrument. The matrix of $a(\theta_i, d_j)$ is often referred to as the optical model.

The data retrieving algorithm is at the heart in light scattering measurements. Due to the oscillatory nature of the Bessel function, which is the mathematical foundation of both the Mie and Fraunhofer formulations, specialized conversion and fitting techniques have to be used. Although there are many matrix conversion and data fitting techniques available in the public domain, the conversion algorithm of commercial instruments is proprietary almost without exception. Besides the optical arrangement, the conversion algorithm is the most important factor in the successfulness of an instrument.

If one has some knowledge about the material, such as whether the size distribution is unimodal or multimodal, narrow or broad, the incorporation of such knowledge into the size resolving procedure may help to obtain a better result since the fitting can now be finely tuned accordingly. For example, certain pretuned fitting conditions can be used and more realistic distributions may be obtained if the size distribution modality is known. However, precaution has to be taken that since such "prechoices" prefix a distribution shape the result will be biased toward the chosen distribution shape. In the other words, you will get the result as you choose. For sophisticated conversion algorithms, such "prechoices" have been proven to be unnecessary and occasionally detrimental in obtaining correct results.

The following two figures display a typical scattering intensity pattern and the corresponding size distribution obtained from a sample with an extremely broad size range (from < 0.1 μm to > 1 mm).

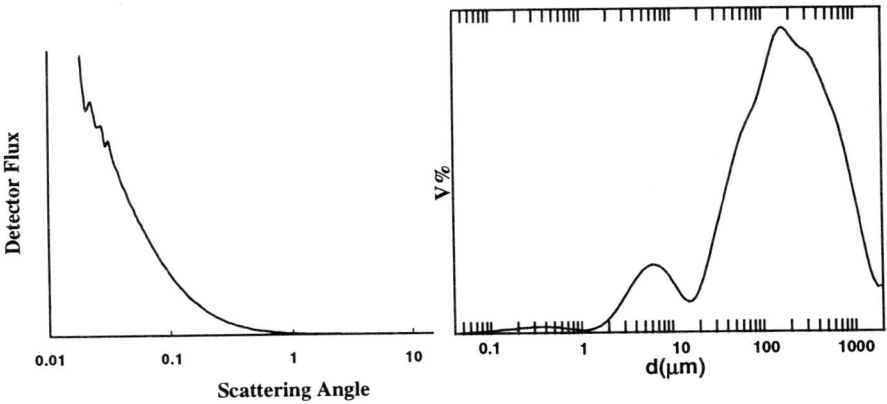

Fig. 3.7: A typical scattering flux pattern measured by Coulter® LS 230 light scattering instrument.

Fig. 3-8: The particle size distribution resolved from the flux in Fig. 3.7.

In obtaining the size distribution from the matrix format, one of the determining factors for obtaining high resolution and correct size distribution is the number of the detectors employed in the detector array. Without enough number of detectors to collect the necessary information from the scattering pattern, even with the best matrix conversion algorithm, high resolution results are not achievable. One can never produce more information from less information obtained: the number of data points in the size distribution should be less or close to, but never more than the number of detectors; i.e., the number of detectors determines the resolution in the size distribution. Fig. 3-9 depicts the resolution and accuracy in finding the minimum position from a detector array with a large number of detectors as compared with that from a detector array with fewer detectors.

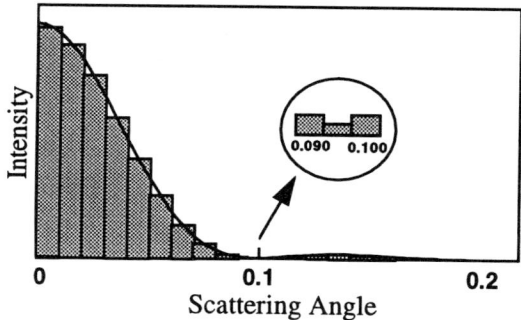

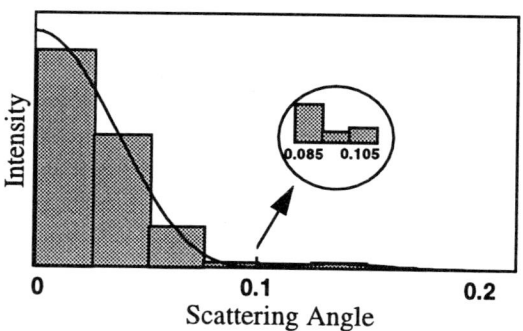

Fig. 3-9: A demonstration of the effect of detector number in size resolution. The upper graph is of a large number of detectors with an angular resolution of 0.01 degree. The detector array used in the lower graph has an angular resolution of 0.02 degree.

In this example, for the array with a large number of detectors, one finds that the first minimum is located between 0.090 and 0.100 degrees which corresponds to one detector angular width and represents the instrument resolution. If Eq. 3-5 is used with $\lambda = 750$ nm, the corresponding size range will be from 582 µm to 524 µm. The same calculation can be performed for the array with the smaller number of detectors. The minimum now is detected between 0.085 and 0.105 degrees. The size range will now be from 617 µm to 499 µm, a much lower resolution. In addition, due to the oscillatory nature of the scattering pattern, if the number of detectors is too small, the fine structures in the scattering pattern will be missed and the ambiguity in the computation will increase resulting in an incorrect size distribution. This is because the computation is based on reported discrete intensity integrated from the detector surface area, not the real continuous angular intensity. In Fig. 3-10, a scattering curve from a mixture of 1:1 volume ratio of 300 μm/600 μm PSLs in water is recorded using a detector array with 126 detectors. If one uses only every fourth point that is equivalent to an array with 32 detectors much of the fine structure in the scattering pattern will be missed. Thus, a correct result resolved from the signal recorded by the 32-detector array will be less likely.

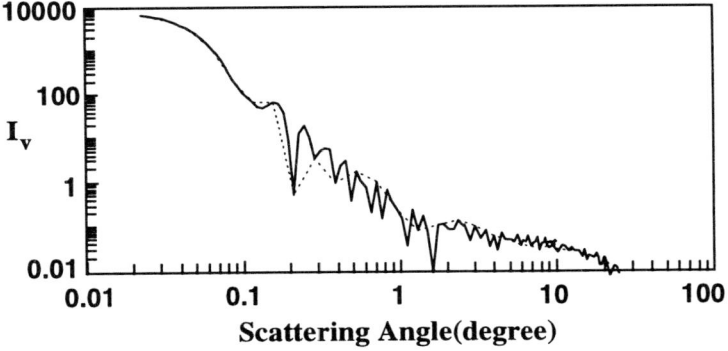

Fig. 3-10: Detected signals from detector arrays of different detector number. Solid line: signal recorded with 126 detectors; dashed line: signal recorded with 32 detectors.

In light scattering measurement, choosing a correct optical model is often the key step in obtaining correct result. Typical errors that may occur when a wrong optical model is chosen (for example, using the Fraunhofer model for particles smaller than 2 μm) are   1) multiple artificial peaks,
2) excessive peak broadening, and
3) incorrect reporting of mean sizes.

3.3 Improvements in Submicron Size Analysis

3.3.1 A Double Lens Optics

It is the central lope (larger particles have a slimmer central scattering intensity peak) along with the fine structures of the angular scattering pattern forming the foundation of resolving particle sizes from light scattering measurements. The upper limit for using a single wavelength of light to characterize particle size by the angular pattern is based on the smallest scattering angle that the instrument can access and partially determined by practical necessity. The upper limit is typically a few millimeters because for particles larger than a few millimeters, other analytical methods, such as sieving, may be more suitable. By applying Eq. 5, the first minimum in the Fraunhofer diffraction for a 3.5 mm diameter spherical particle, for example, is located at $\theta = 0.015$ degrees. Thus, in order to detect and to fit the sharp intensity change within the central lobe, the smallest scattering angle accessible should be at least smaller than 0.01 degrees. Typically, a very long optical bench is needed to reach that range.

The lower limit is determined when the fine structure of the angular pattern diminishes in the detecting angular range. The value for the lower limit depends on the relative refractive index and the optical arrangement. Conventionally, the optics in a light scattering instrument is composed of a light source, a sample chamber, a Fourier lens and a detector array as the portion inside the dot line in Fig. 3-11.

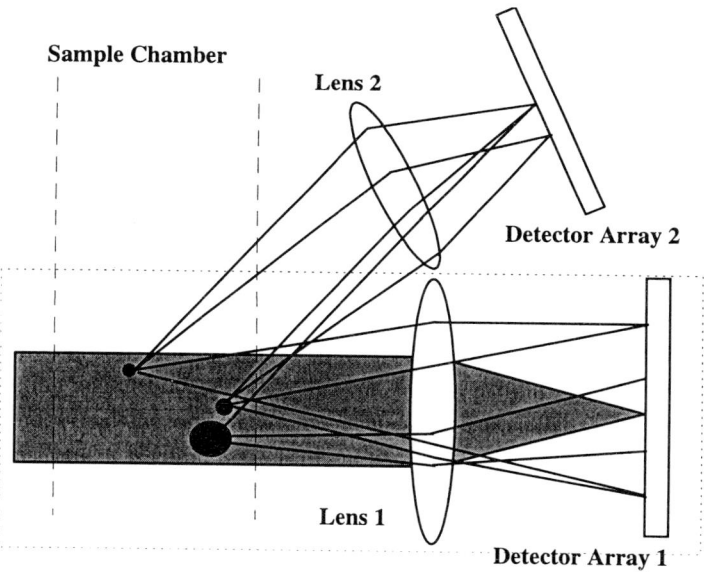

Fig. 3-11: A double lens optical system.

Particles in the sample cell scatter light from the incident beam. Upon leaving the sample cell, the scattered light is guided to the detector plane by the lens. Particles pass through an expanded and collimated laser beam in front of a lens in whose focal plane is positioned a photosensitive detector array. The focal length of the lens, the detector array arrangement and the laser wavelength together determine the particle size range that can be analyzed. The main incident beam is brought to focus in the detector plane and either reflected out by a mirror or passed through an aperture at the detector center.

The angular scattering pattern is recorded by the detector array at the detector plane. This light intensity function is commonly refereed to as the primary data, a light flux pattern f(θ), where flux is defined as light intensity per unit detecting area. Light scattered at a specific angle will be directed by the lens to fall onto a particular detector regardless of the particle's position. Particles may flow or move through any part of the beam but their diffraction pattern is stationary and proportional in intensity to the total number of particles contributing to the scattering. For a polydisperse sample, the angular pattern may change with time as the number distribution of particles in the scattering volume varies over time with particles entering and leaving the beam. This temporal variation is usually integrated to give an average true representation of the sample.

If one wants to size particles by using the characteristic maxima and minima of the intensity, in order to measure smaller particles a wider detecting angle is needed because the scattering pattern becomes flatter with maxima and minima spreading over a wider angular range as shown in Fig. 3-3. In typical optical setups, the focusing lens has a finite size so that the scattering angular range from which scattered intensity can be collected by a single lens is limited typically to less than 15-20 degrees. One obvious way is to increase the dimension of the focusing lens and the size of the detector array. However, besides the cost and size of expensive large lenses, the quality of lens grinding and associated aberrations that affect high angle signals are more severe as the lens diameter gets bigger. Another alternative is to use multiple lasers positioned in different locations so that the same detector array will receive the scattering from the same particle but illuminated by different lasers in different scattering angular ranges. However, alignment of multiple lasers, for both light intensity and laser position, is inherently complex and may be unreliable. The sample cell design is also exceedingly complex. In addition, the intensity and polarization fluctuations of the three lasers will greatly affect the uniformity of the angular pattern leading to possible additional measurement error because the angular pattern is composed of three consecutive independent measurements. A better way to achieve a wider angular coverage is to use a second lens and a second detector array placed at a high angle to extend the detecting areas, as is shown in Fig. 3-11.

The addition of the second lens and second detecting array extends the detecting angle to more than 35 degrees which effectively extends the lower sizing range down to about a few hundreds of nanometers [8].

3.3.2 Polarization Intensity Differential Scattering (PIDS)

Currently, light scattering technology is being applied to an ever widening variety of fields. Many samples have particles whose sizes extend into the submicron range, which requires instruments capable of sizing particles in wider size ranges. However, when particle size gets smaller, the scattering pattern becomes smoother and less angular dependent. Thus, it is more difficult to obtain correct sizes from forward angle scattering alone, even with the double lens optics that covers the scattering angle up to 35 degrees.

To further enhance the sizing capability when utilizing the scattering angular pattern, there are three approaches one may use. When combining these three approaches, the theoretical limit in sizing particles by light scattering technology may be approached.

The first step used to increase the sizing range is to increase the detecting angular range. For large particles, their characteristic patterns are mainly located at small angles. When particles get smaller, the central lobe gets flatter and fatter and the angular pattern gets less characteristic. If we use the angle location of the first minimum in the scattering pattern as the necessary detection coverage to correctly size a sphere, we found that in order to size a sphere with the diameter less than 0.5 um, the maximum detecting angle has to be larger than 90 degrees. Thus, to size a submicron particle, the detecting angular range has to be increased to cover angles larger than 90 degrees, at least.

The scattering patterns are a function of light wavelength and particle size and their variation is related to the ratio of particle dimension and light wavelength (d/λ), which can be used as an indicator for the effectiveness in using scattering angular pattern to size particles. Interference effects are greatly reduced when d/λ is less than 0.5. Obviously, if the wavelength of light is shorter, the ratio will be greater and the lower sizing limit will be effectively extended. In other words, the shorter wavelengths compress the scattering patterns so that more information (the fine structures in the scattering patterns) can be obtained in the same angular range when compared with what is obtained using a longer wavelength of light. The practical shortest wavelength would be ca. 350 nm because most materials have strong absorption when the wavelength is shorter than 300 nm. By using a light of 375 nm, the lower sizing limit can be further extended to twice lower when compared with using a laser light of 750 nm. For example, in Fig. 5, for a 0.5 μm PSL sphere, when a wavelength of $\lambda_o = 450$ nm is used the fine structure is more distinguishable when compared with that in using a longer wavelength ($\lambda_o = 750$ nm).

The above two approaches used to increase the size coverage can be quantitatively described by the magnitude of the momentum-transfer vector K (Eq. 3-1). For large particles because the intensity changes rapidly with the scattering angle, the overall K range covered by the detectors should be small in order to detect the details and reach a high resolution. For small particle, it is just the opposite that the detectors should cover a K range as big as possible. For example, if one uses a laser light ($\lambda = 750$ nm) to measure scattering up to 35 degrees, the $K_{max} = 0.005$ nm^{-1}. The K_{max} will be increased more than five times to 0.027 nm^{-1} with a light of $\lambda = 450$ nm at a wider angular range up to 150 degrees.

However, even with the above two approaches, the realistic lower limit in sizing is still not low enough to meet the demand for sizing even smaller particles (tens of nanometers in diameter). When particles get even smaller, further increase in the scattering angle will not be significant because of the even slower angular variation for smaller particles.Fig. 3-12 is a zoomed 3-D display from Fig. 3-3 that displays the very slow angular variation for small particles. In this case, once the size is smaller than around 0.4 μm not only may the intensity minimum disappear but also will the maximum intensity contrast in the entire angular range become so small that prevent such angular pattern from to be used in obtaining size information. Fig. 3-13 shows the maximum contrast in the scattering intensity I_V from 0.5 degrees to 150 degrees at $\lambda = 450$ nm for 100 nm spheres of different refractive indices. Fig. 3-14 displays the maximum intensity contrast for submicron particles in the scattering angular range from 0.5 degrees to 175 degrees at two different wavelengths. As shown in Fig. 3-14, for particles less than 200 nm even at $\lambda = 450$ nm the maximum intensity contrast is too small to effectively retrieve correct particle sizes. This finding demonstrates that for particles smaller than 200 nm, even taking advantages of the above two approaches, e.g., using higher angles and shorter wavelengths, it is still impossible to obtain correct sizes.

Characteristic of I_H for small particles is that there is a minimum at around 90 degrees. This minimum will shift to larger angles for larger particles. Thus, although both I_V and I_H have small contrast for small particles, their difference shows more distinguished fine structures that make sizing small particles possible. Combining the polarization effect of light scattering with the wavelength dependence at high angles, we can extend the lower sizing limit to as low as 50 nm, almost reaching the theoretical limit. This is the patented polarization intensity differential scattering (PIDS) technique [9,10].

Scattered light at horizontal polarization has different scattering patterns and fine structures from that of vertically polarized light for small particles. The main

The origin of the polarization effect can be understood in the following way. If a very tiny particle, which is much smaller than the light wavelength, is located in a light beam, the oscillating electric field of the light induces an oscillating dipole moment in the particle, i.e., the electrons in the atoms comprising the particle move back and forth relative to the stationary particle. The induced motion of the electrons will be in the direction of oscillation of the electric field, and therefore perpendicular to the direction of propagation of the light beam. As a result of the transverse nature of light, the oscillating dipole radiates light in all directions except in the direction of oscillation as visualized in Fig. 3-15. Thus, referring to Fig. 3-1, if the detector is facing the direction

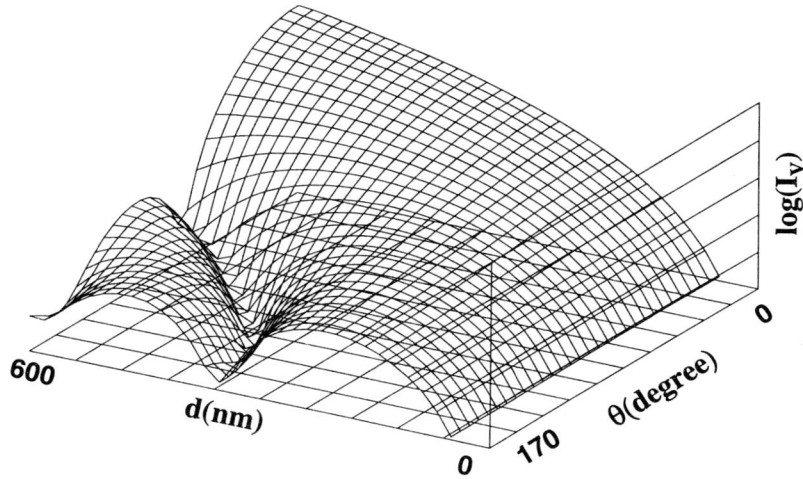

Fig. 3-12: A 3-D display of the Mie scattering intensity I_V from a unit volume of spheres with the relative refractive index m = 1.50 +0i. The light wavelength is 750 nm.

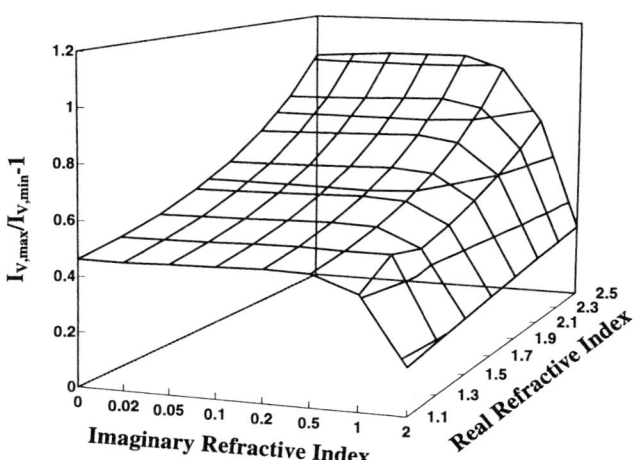

Fig. 3-13: Maximum scattering intensity I_V contrast from 0.5 degrees to 150 degrees of scattering angles for 100 nm spheres of different refractive indices at λ = 450 nm.

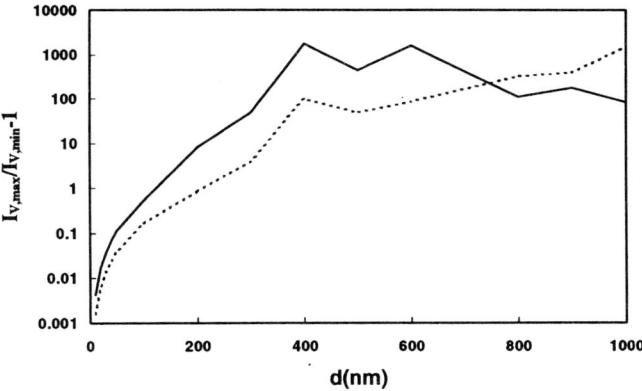

Fig. 3-14: Maximum scattering intensity I_V contrast in the scattering angular pattern from 0.5 degrees to 150 degrees for spheres (m = 1.5-0.02i) in air at the wavelengths of λ = 450 nm (the solid line) and λ = 750 nm (the dot line).

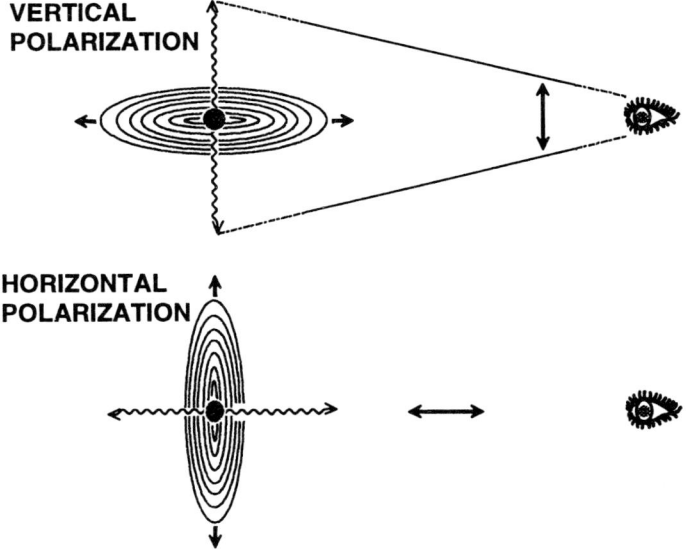

Fig. 3-15: Scattering from different polarizations.

of oscillation it will receive no scattering from single dipoles. When the light beam is oriented in either V direction or H direction, the detected scattering intensity I_V and I_H at a given angle will be different. The difference between I_V and I_H (I_V-I_H) is termed the PIDS signal.

As the particle size increases the intra-particle interference makes the particle's behavior deviated from a simple dipole and the scattering pattern will become more complex. For small particles the PIDS signal is roughly a quadratic curve centered at 90 degrees. For larger particles, the pattern shifts to smaller angles and secondary peaks appear. Since the PIDS signal is dependent on particle size relative to the light wavelength, valuable information about the particle size distribution can be obtained by measuring the PIDS signal at a variety of wavelengths. In order to visualize the polarization effect in the scattering intensity, let us look the difference in I_V and I_H as a function of diameter and angle for particles of relative refractive index being 1.20 at $\lambda_o = 450$ nm.

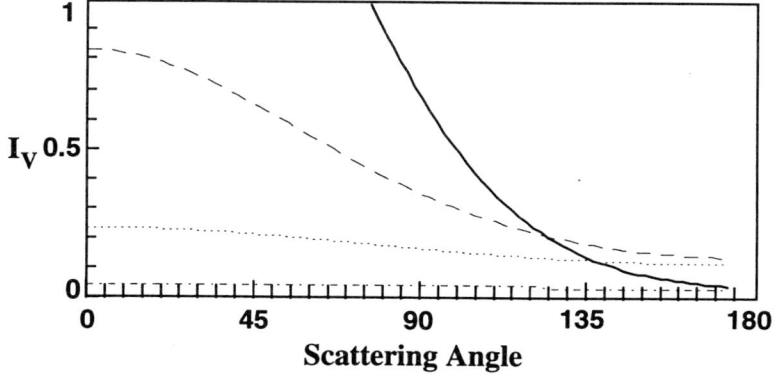

Fig. 3-16: I_V scattering of small PSL in water ($\lambda_o = 450$ nm). Solid line: d = 200 nm; dashed line: d = 150 nm; dotted line: d = 100 nm; and dot dashed line: d = 50 nm.

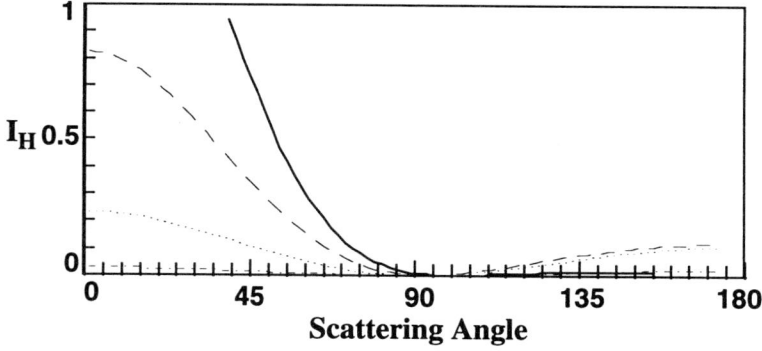

Fig. 3-17: I_H scattering of small PSL in water ($\lambda_o = 450$ nm) with the same symbols as that in Fig. 3-16.

The main difference between the plots in Fig. 3-16 and in Fig. 3-17 is that there is a minimum around 90 degrees in the I_H scattering for each size. Thus, the angular variation of the scattering intensity for small particles can be enhanced if the differential intensity between I_V and I_H (=I_V-I_H) is used instead of I_V as illustrated by Fig. 3-18. In addition, because the PIDS signals vary at different wavelengths (it gets flatter at longer light wavelengths), measurement of PIDS signals at several wavelengths will provide additional scattering information that can be used to refine the size retrieving process.

Fig. 3-19 is a plot that displays the maximum PIDS contrast in the entire angular range. Comparing with Fig. 14, it can be clearly seen that the improvements for particles smaller less than 300 nm is significant. From Fig. 18, the angular patterns for 100 nm

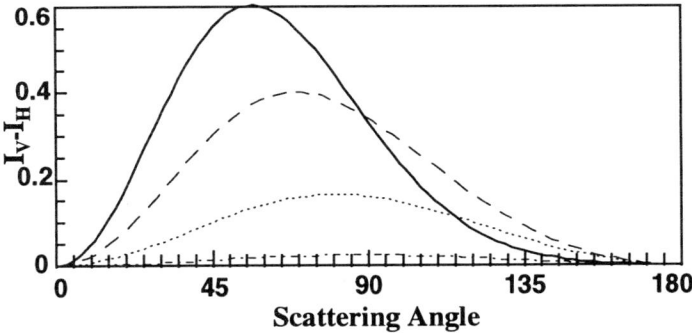

Fig. 3-18: I_V-I_H scattering of small PSL in water (λ_o = 450 nm) with the same symbols as that in Fig. 3-16.

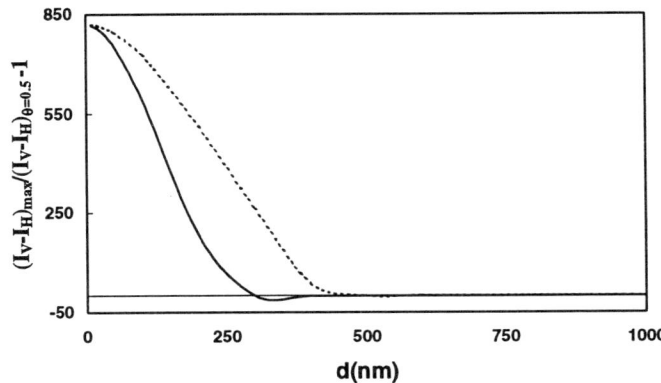

Fig. 3-19: Maximum PIDS contrast in the PIDS angular pattern from 0.5 degrees to 150 degrees for spheres (m = 1.5-0.02i) in air at the wavelengths of λ = 450 nm (the solid line) and λ = 750 nm (the dot line).

and even for 50 nm particles are recognizable, in addition to the shift in the axis of symmetry. It has been verified through both theoretical simulation and real experimentation that to accurately size particles smaller than approximately 0.4 μm by scattering intensity without the PIDS technique is just unrealistic and cannot be true. The combination of the three approaches, wider angular range, wavelength variation, and polarization effect, especially the last one, improves the capability of characterization of submicron particles using light scattering angular pattern. One additional feature of the PIDS technique is that because PIDS measures the difference between two intensity measurements which are time-averaged and detected by the same detector, the noise (to the first order) is minimized.

3.4 PIDS Instrumentation and Data Analysis

3.4.1 Optical Setup For PIDS

In modern light scattering instrumentation a laser light, either a He-Ne gas laser ($\lambda_0 = 632.8$ nm) or a diode laser (λ_0 ranges from 650 nm to 780 nm), is always the main light source or the sole light source because of its high intensity, high stability, long lifetime, and polarized and monochromatic illumination. However, for PIDS measurement, in order to obtain a light with several wavelengths at different polarizations, for which a laser light is not suitable, a white light source, typically a tungsten-halogen lamp, has to be used. A tungsten-halogen lamp has random polarization and a broad spectrum range with the wavelength typically ranging from 250 nm to 3000 nm. By placing different wavelength filters and polarizers in front of the light beam before the sample chamber one can make light scattering measurement at selective wavelengths and different polarizations.

In a typical PIDS optics system (Fig. 3-20), light from a white light source is filtered through filters of several wavelengths at two polarizations. The collimated light passing through the sample cell is monitored by a beam monitor. The scattered light is detected by detectors located over a wide angular range, typically from 60 degrees to 145 degrees. As sample is introduced to the sample cell, light begins to scatter from its direct path along the optic axis, and the signal in the beam dump decreases. When there is sufficient sample in the sample cell to provide a good signal to the photodetectors, yet not so much as to cause multiple scattering, the sample measurement begins. During the measurement the response from each of the detector elements is amplified by a dedicated amplifier circuit, and this signal is averaged over the duration of the measurement, typically one minute or longer. When the measurement is completed, these averaged values made at each wavelength and

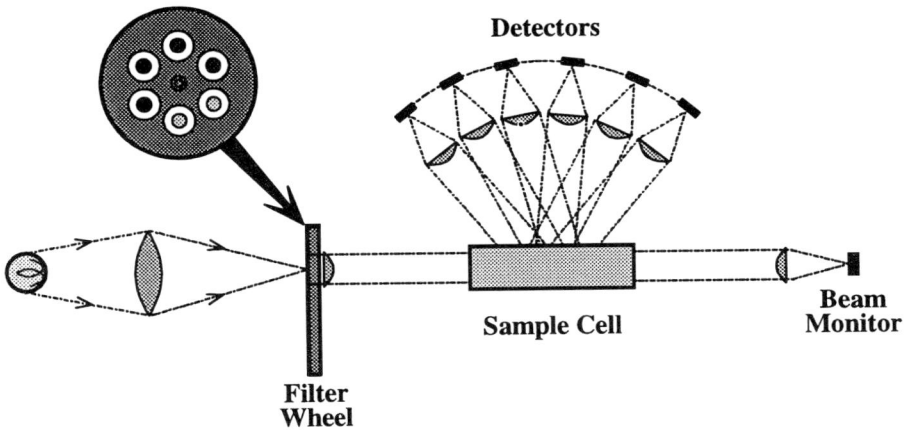

Fig. 3-20: A typical PIDS system that utilizes multiple wavelengths and polarizations to size submicron particles.

polarization sequentially are written to a data file and then combined with the data from the forward scattering measurement providing a light scattering instrument the capability of sizing submicron particles as small as some tens of nanometers.

3.4.2 Data Analysis

The data analysis procedure for the scattering experiments combining low angle scattering with high angle PIDS measurements is similar to the procedure described earlier. The matrix form is now extended to include both forward scattering and PIDS information since both forward scattering and PIDS signals are mathematically based on the same theory and physically measured from the same material in the sample chamber. In Eq. 7, $p(\theta_i,\lambda_j)$ is the PIDS signal measured at angle θ_i (i= i to k) and at wavelength λ_j (j=1 to h). Since the light sources and the detectors are different for low angle scattering and PIDS measurement, the signals from them have to be weighted differently before they can be combined into a single matrix. The weighting and scaling constants can be easily obtained from an instrument calibration procedure.

$$\begin{bmatrix} f(\theta_1) \\ ... \\ f(\theta_m) \\ p(\theta_i,\lambda_1) \\ ... \\ p(\theta_k,\lambda_h) \end{bmatrix} = \begin{bmatrix} a(\theta_1,d_1) & ... & ... & a(\theta_1,d_n) \\ ... & ... & ... & ... \\ a(\theta_m,d_1) & ... & ... & a(\theta_m,d_n) \\ b(\theta_i,\lambda_1,d_1) & ... & ... & b(\theta_i,\lambda_1,d_n) \\ ... & ... & ... & ... \\ b(\theta_k,\lambda_h,d_1) & ... & ... & b(\theta_k,\lambda_h,d_n) \end{bmatrix} \begin{bmatrix} x(d_1) \\ ... \\ x(d_n) \end{bmatrix}$$

Eq. 3-7

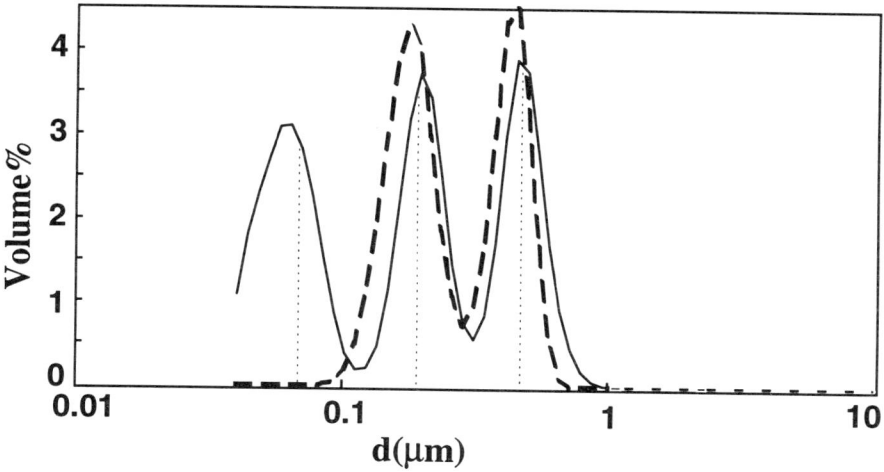

Fig. 3-21: A real trimodal mixture of PSL (1:1:1 volume ratio) measured by Coulter® LS 230 light scattering instruments using PIDS technique (solid line) and without using PIDS (dashed line). The dotted lines are the nominal diameter values of the PSL in the mixture from the PSL vendor.

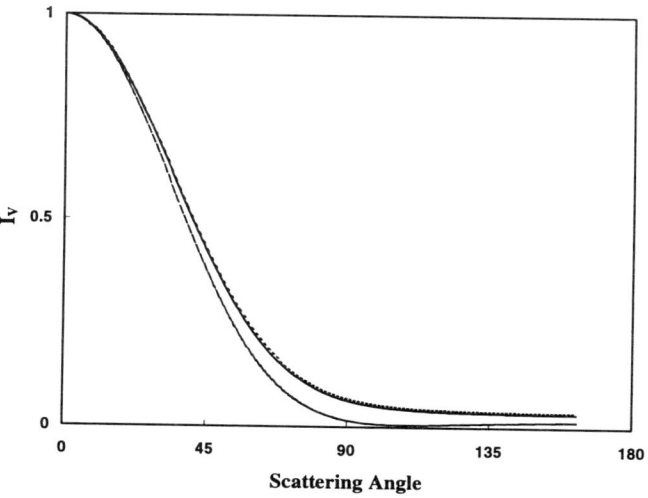

Fig. 3-22: Normalized I_V scattering pattern from mixtures of PSL spheres in water at λ_o = 750 nm. The sphere diameters of each component obey the log-normal distribution with a variance of 0.01. Dashed line: 460 nm spheres; solid line: 1:1 mixture of 240 nm and 460 nm spheres; dotted line: 3:1:1 mixture of 64 nm, 240 nm and 460 nm spheres.

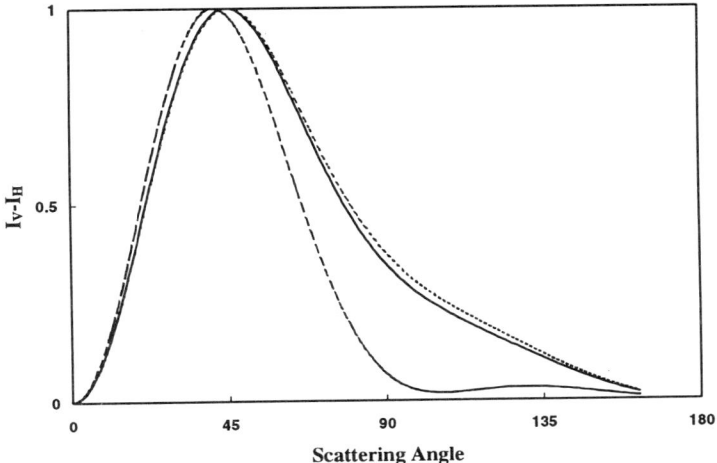

Fig. 3-23: Normalized PIDS pattern from the same mixtures of spheres as in Fig. 3-22. The symbols are the same as these in Fig. 3-22.

Fig. 3-21 shows a typical trimodal distribution with the diameter ratios being approximately 1:2:4. Without the PIDS technique the smallest component is missed, even using the information gathered from high scattering angles and short wavelengths. Let us do a signal analysis from the simulated data of the same sample as that in Fig. 3-21 to see why the PIDS technique can provide high resolution to resolve all three peaks while using even the short wavelengths and high angles the peak for the smallest size is still missed.

We start with a log-normal size distribution of PSL spheres in water, sample A. The distribution has a mean diameter of 460 nm and a variance of 0.01. Using Eq. 3-6 we calculate the I_v scattering angular pattern for the spheres at $\lambda = 563$ nm (the dashed line in Fig. 3-22). Then we mix a log-normal size distribution of PSL having a mean diameter of 240 nm and a variance of 0.01 with sample A at a mixing ratio of 1:1 to form sample B. The corresponding I_v scattering pattern now is different from that of sample A (the solid line in Fig. 3-22). When we add a third component (a log-normal size distribution of PSL having a mean diameter of 64 nm and a variance of 0.01) with sample B at a mixing ratio 3:1 to form sample C, we do not see any appreciable change in the scattering pattern (the dotted line in Fig. 3-22). The size retrieving process from measurement is the matrix inversion process of the above computation. One should be able to resolve the difference between sample A and sample B, i.e., separating the 240 nm peak from the 460 nm peak from a careful measurement using a sophisticated instrument. However, as there is only minimal change in the I_v pattern from sample B to sample C, one cannot expect to resolve the added small particles (the 64 nm spheres) from the large particles within the experimental error limit, no matter how high the signal-to-noise ratio and how smart the fitting algorithm are.

On the other hand, when we do the same simulation and comparison for the PIDS signals from the above three samples, not only does the PIDS patterns from sample A and sample B show a bigger difference than the I_v patterns for the samples, the PIDS pattern from sample C is also easily distinguishable from that of sample B. The difference between the PIDS patterns of sample B and sample C provides a foundation for resolving the 64 nm peak in sample C from light scattering measurement using the PIDS technique as it has been experimentally verified by Fig. 3-21. Fig. 3-24 shows another example of using the PIDS technique to recover size information of small particles.

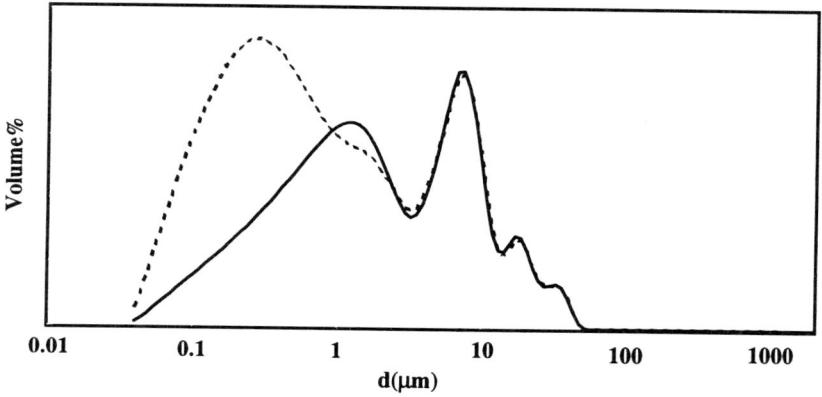

Fig. 3-24: The size distribution of an industrial sample measured by Coulter® LS 230 instrument with (the dashed line) and without (the solid line) using the PIDS technique.

3.5 Summary

In summary, because of the extremely wide dynamic range and fast measurement associated with the technology, elastic light scattering intensity measurement has become one of the most popular methods in sizing particles from nanometers to millimeters -- replacing many other existing sizing technologies such as sieving, sedimentation and microscopy in a broad range of industrial fields [for examples, 11-13]. Diffraction measurements lose resolution when the particle size is smaller than the wavelength of light (Fig. 3-14). In contrast, the PIDS technique has the highest sensitivity, highest resolution for particles smaller than the wavelength (Fig. 3-19) [14]. The combination of these two complementary techniques: the PIDS technique and conventional forward scattering method, makes the characterization of particulate systems with broad distributions span from 50 nm to 2 mm possible.

3.6 Appendix

Notation

$a(\theta,d)$	element of scattering kernel	m	m_1/n_0
$b(\theta,\lambda,d)$	element of PIDS kernel	m_1	refractive index of particle (may be a complex number)
d	particle diameter	n_0	refractive index of medium
E_V	scattered electric field with vertically polarized light	$p(\theta,\lambda)$	PIDS signal at angle θ and wavelength λ
E_H	scattered electric field with horizontally polarized light	$x(d)$	size distribution
		α	$\pi d n_0/\lambda_0$
$f(\theta)$	intensity flux at angle θ	β	$\pi d m/\lambda_0$
I_V	scattered intensity with vertically polarized light	λ_0	wavelength of light in vacou
		λ	wavelength of light in medium
I_H	scattered intensity with horizontally polarized light	θ	scattering angle
		φ	angle between Z-axis and scattered light (Fig. 3-1)
J_1	Bessel function of the first kind of order unity		
K	momentum-transfer vector		
K_0	incident wave vector		
K_s	scattered wave vector		

3.7 References

1. G. Mie, Ann. Physik., 25: 377. (1908).

2. S. Asano and M. Sato, Appl. Optics, 19: 962. (1980).

3. R. J. Martin, J. Modern Optics, 40(12): 2467. (1993).

4. M. Kerker, The Scattering of Light and Other Electromagnetic Radiation, Academic Press, New York, (1965).

5. J. Komrska, Optica Acta, 19(10): 807. (1972).

6. J. Komrska, Optica Acta, 20(7): 549. (1973).

7. K. A. Kusters, J. G. Wijers and D. Thoenes, Appl. Opt. 30(33): 4839. (1991).

8. S. E. Bott and W. H. Hart, US Patent 5056918, (1991).

9. S. E. Bott and W. H. Hart, ACS Symposium Series 472: Particle Size Distribution II, (T. Provder, ed.), American Chemical Society, Washington DC, p. 106. (1991).

10. S. E. Bott and W. H. Hart, US Patents 4953978, (1990); 5104221, (1992).

11. J. L. Loizeau, D. Arbouille, S. Santiago and J. P. Vernet, Sedimentology, 41: 353 (1994).

12. F. W. C. den Ouden and T. van Vliet, Special Publication, Royal Soc. Chem, 113: 285 (1993).

13. G. A. Hareland, J. Cereal Science, 21: 183 (1994).

14. G. Row, Submicron Particle Sizing Using Laser Diffraction Instruments, Coulter Technical Monograph, Coulter Corp., Miami, Florida (1993).

Address of the author:
Renliang Xu
Coulter Corporation
Miami
Florida

4 Particle Sizing in the Sub-micron Range by Laser Diffraction

M. W. Wedd, Malvern Instruments Limited, Malvern

4.1 Introduction

Particle size analyzers using Low Angle Light Scattering have become a respected and popular technique in the past few years.

The reason for this is sound as they are easy to use, do not require calibration in the traditional sense and supply a rapid and highly reproducible answer. All of these instruments measure the light scattered from a system of particles via an optical arrangement to a series of detectors which record a current proportional to the intensity of the scattered light falling upon them.

We obtain from the measurement a distribution by volume of a range of particle sizes which covers the sample measured. To make the link between the light scattered and recorded in the computer we need to be able to accurately predict how various sizes of particles will scatter light and in what proportion does the intensity of this scattered light relate to the volumetric distribution of particle sizes present.

To do this we need both a mathematical model of the way that particles scatter light as well as a numerical inversion routine which transposes the scattered light to a distribution of particle sizes by volume.

The uncertainties of this technique stem from inappropriate selection of which model of scattering is to be employed coupled with sampling and dispersion problems. Wet dispersion, dry powders and sprays are measured. The technique is now starting to be used on-line.

For single particles the general scattered intensity can be written as

$$I_{SCAT} = \frac{I_\circ}{K^2} \int_0^{2\pi} \int_0^{\pi} F(\theta\varnothing) \sin\theta \, d\theta \, d\varnothing \qquad \text{Eq. 4-1}$$

$I_\circ$ is the incident intensity and $K = \frac{2\pi}{\lambda}$

λ is wavelength of $I_\circ$ of the illuminating source in the media.

$F(\theta\varnothing)$ describes the radiance from a particle at the center of a spherical ordinate system.

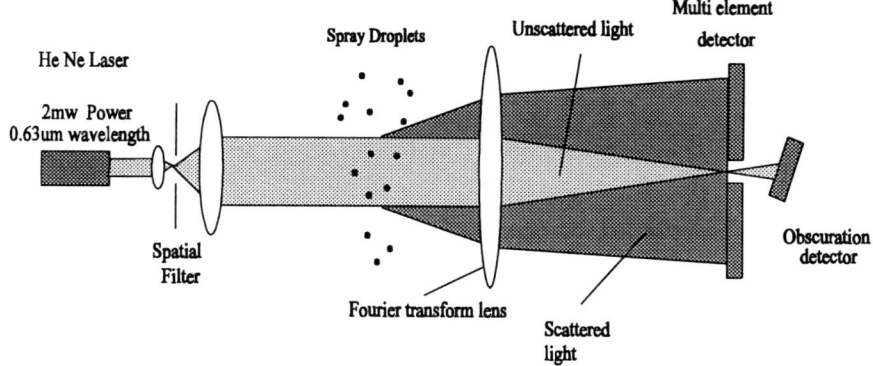

Fig. 4-1: Components of a laser diffraction particle sizer

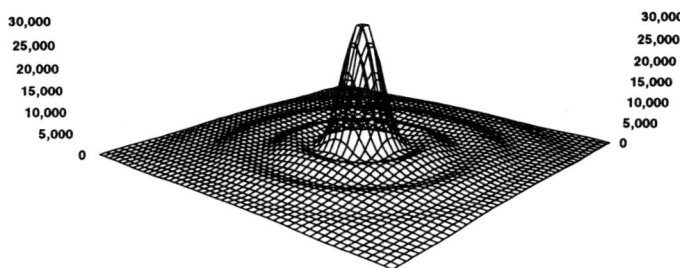

Fig. 4-2: The intensity of a 3 µ latex particle RI = 1.6 - 0.0, wavelength 633 nm

Historically the approximation of Fraunhofer was the first model employed. Fraunhofer assumptions for individual particles are:

a) That all the particles are totally opaque.
b) That spheres will yields the same scattering pattern as a thin two dimensional circular disk.
c) That the extinction efficiency is 2 for all particle sizes (i.e. it does not predict that for small particles the intensity of scattering falls more rapidly than the geometric cross section would predict).

For the Fraunhofer approximation the equation can be written as:

$$I \, \alpha \left[\frac{2J_1(Ka\sin\theta)}{Ka\sin\theta} \right]^2 \qquad \text{Eq. 4-2}$$

where J_1 is Bessell function of the first kind order 1. θ angle of scattering
a = particle size & $k = \frac{2\pi}{\lambda}$ where λ is the illuminating wavelength.

The approximation of Fraunhofer does not make use of any knowledge of the optical properties of the material. In practice this is a valid assumption for large particles having a high relative refractive index compared to the media they are suspended in. However, for small particles, even if they are opaque, which have a low relative refractive index, errors in the proportion of volume subscribed to a given size occur. The equations of Fraunhofer are relatively simple and quick to calculate.

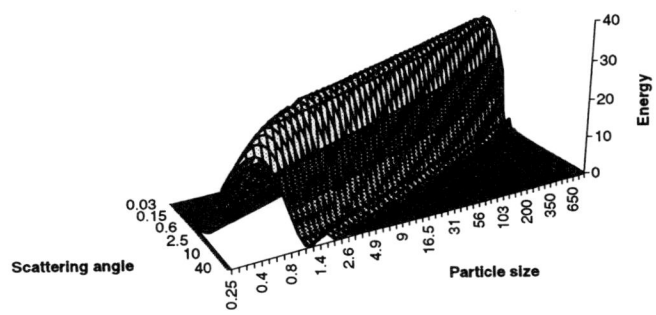

Fig. 4-3: Light energy plotted against particle size and scattering angle for equal volumes of particles Fraunhofer

4.1.1 Anomalous Diffraction Approximation

This model also has limited application and assumes:-
That all the particles are spheres
That all the particles are totally transparent
That the difference in refractive index between the particles and the media is small (<1.1) and known. As this approximation does include an estimate of how the extinction efficiency varies with particle size and thus does predict to a degree that variation of scattering efficiency with size its use in circumstances of particles suspended in liquids did result in more authentic size distributions in many cases.
Equations having similar characteristics but of increasingly more complex form can be written for Lorenz-Mie theory.

4.1.2 Lorenz-Mie Theory

The advent of reasonably powerful desk top computers enabled the rigorous solution to the scattering of homogeneous spheres to be coded and calculated in reasonable times. by this means some of the uncertainties of the previous models have been eliminated. However, to exploit this theory to the full a knowledge of the optical properties of the particles and the suspending media are needed in terms of a value for the refractive index.

The theoretical predictions of the scattering of particles are then used in conjunction with the geometry of the chosen detector to form a "T-matrix" which describes how unit volumes of each size of particle would appear as energy to each detector element. The example shown clearly indicated shows the weak contribution made by sub 0.25 micron particles.

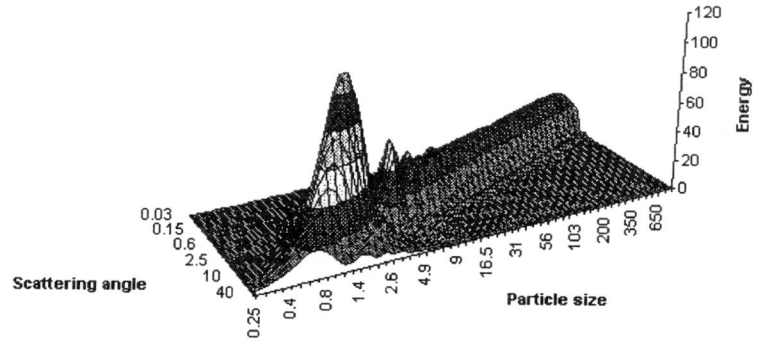

Fig. 4-4: Light energy plotted against particle size and scattering angle for equal volumes of particles RI = 1.6 - 0.0 in water

The light scattered by single spherical particles can be extended to clouds provided:
Each particle scatters as an independent entity (i.e. no significant multiple scattering).
That there is no optical interference between the scattered radiation from different particles.
This is satisfied if all the particles move randomly with respect to each other and that the overall scattering pattern is sampled many times.

4.1.3 General Lessons of Lorenz-Mie Theory

The theory predicts that various sizes of spherical particle exhibit an angular scattering dependence, an amplitude of scattering dependence and a polarization dependence.

Polarization Properties

As the intensity of the light scattered by particles less than 0.3 micron reduces so quickly with further reduction in particle size, it becomes increasingly more difficult to discriminate their scattering contribution in the presence of the larger particles.
Inspection of Fig. 4-5 shows that at an observed angle of 90 degrees the difference in scattering between two orthogonal planes of polarization (the dotted, as compared with the solid curves), appears to increase as the particles become smaller.

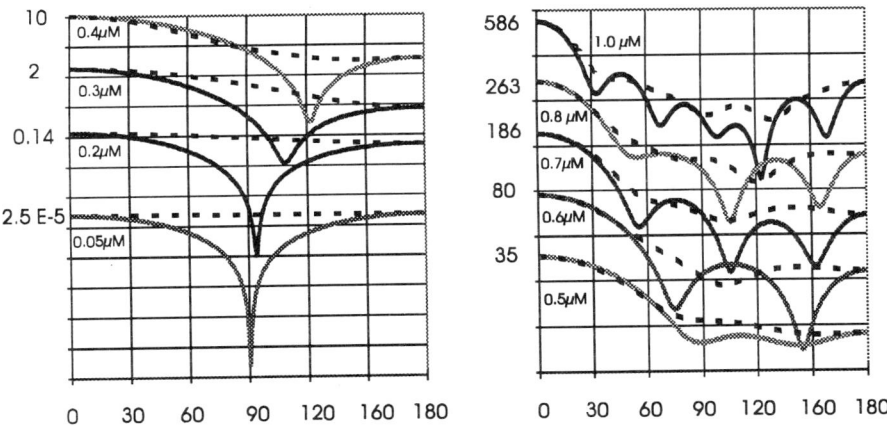

Fig. 4-5: Scattering of single non-absorbing spheres

This feature is exploited in the so called Polarization Intensity Difference Scattering (PIDS), Fig. 4-6. The amplitude of this difference also changes with the wavelength of illumination.

In most implementations of this approach a white light source is filtered to provide 3 or more wavelengths in the visible light band, typically 450 nm 600 nm and 900 nm. At each wavelength the intensity is recorded at two detectors, one positioned such that its polarization has its electric vector perpendicular to the plane of the incident light and a second detector parallel to that plane. From the six data values the volumes of material at three size points, typically 0.15, 0.20 and 0.30 micron are obtained. These additional data points are added to low angle light scattering result by some heuristic blending routine.

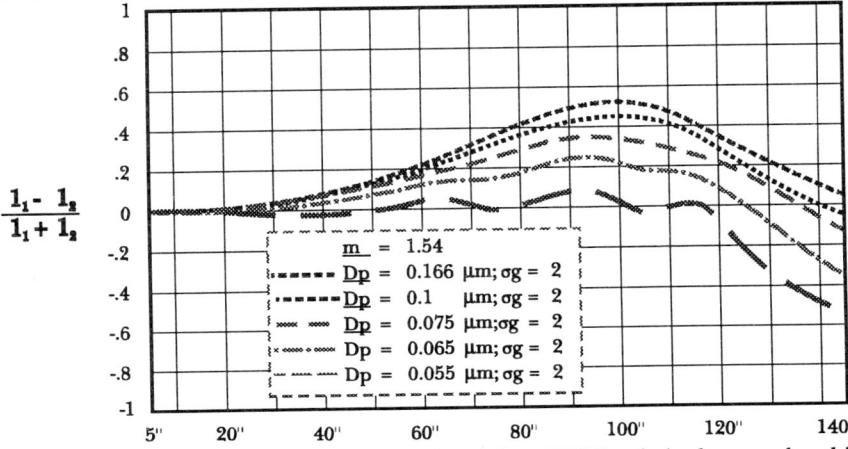

Fig. 4-6: Polarization Intensity Difference Scattering (PIDS) of single non-absorbing spheres

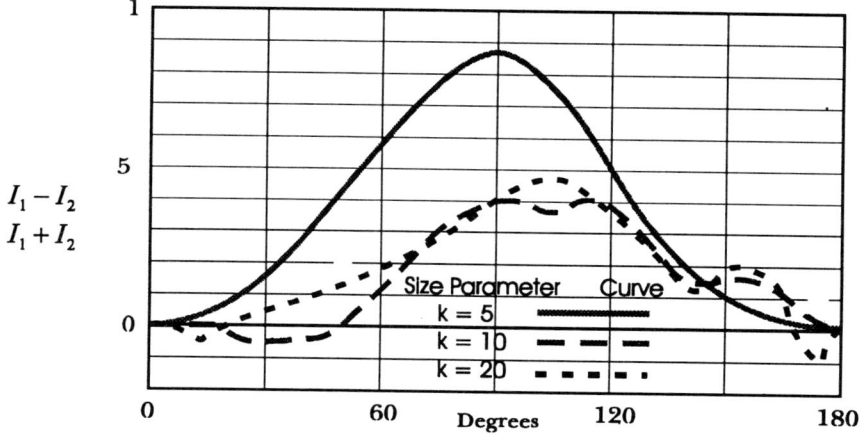

Fig. 4-7: PIDS of an oblate spheroid, (Length / Width) = 5 $k = \dfrac{2\pi F(a)}{\lambda}$

Any non-spherical particle will exhibit a degree of polarization to the light scattered from it in addition to that specifically due to sizes less than the illuminating wavelength. Thus a further problem arises in that large irregular particles will excite the polarization sensing detectors leading to the false conclusion that sub-micron particles are present when no such real material of that size exists in the sample.

Fig. 4-7 shows that when both size and shape are present together, as is the case for the Oblate spheroid, then ambiguous results are obtained. In the example shown in Fig Fig. 4-7, particles whose size differs by a factor of two yields the same Polarization Intensity Difference at angles around 90 degrees.

Studies of the fine particles in the upper atmosphere have been examined using polarization, but specific extra tests must be conducted to eliminate the possibility of ice crystals being present which are irregular and usually larger than the targeted particles under study and are well known to totally distort the measurement. (These experiments also have the added advantage of using the sun which emits a wider spectrum of wavelengths thus extending the size range possibilities).

In general, sweeping simplifications about the scattering of particles is to be avoided. The interrelated dependence between the particle size, real refractive index, and absorbence (imaginary refractive index) can be slightly simplified by consideration that the particle size, real refractive index, and absorbence (imaginary refractive index) are three mutually perpendicular vectors with the following distinctive values.

- The particle size is small (less that 10 microns).
- The relative Refractive Index (real) is small (less that 1.2).
- The absorbence (imaginary Refractive Index) is small (less that 0.05).

Then if any value of two of the above parameters is large, then the influence of the third is weak. In regions where one or two of the parameters is small, good knowledge of the optical properties is essential.

An important requirement of the use of Lorenz-Mie theory is the need to know the Refractive Index of the particles being measured.

4.2 What is Refractive Index?

Light is an electromagnetic wave in just the same way as television and radio waves are.

The waves are characterized by having an electrical intensity vector with a magnetic intensity vector at ninety degrees to it.

If the electric vector is up-down the light can be said to be vertically polarized. (i.e. the magnetic vector is horizontal).

Both vectors are varying in intensity in a sinusoidal manner at some frequency. For HeNe laser light this results in a wavelength of $\lambda = 632.8$ nm

The resultant wave interaction with particles is complex.

In order to describe how a particle will interact with both the electric and the magnetic vector a similar complex vector "refractive index" has been devised.

Vector interactions are describable by complex arithmetic of the style $A + jB$ where A and B are vectors of magnitude A & B at ninety degrees to each other.

In practice the two values of the real refractive index (A) and the imaginary refractive index (B) describe the way the material interacts with the electric vector (A) and the magnetic vector (B).

For gold or other non oxidizing conductive metals the imaginary refractive index interacts strongly with the magnetic vector and indeed induces electrical currents to flow within the particle. For gold the Refractive Index = 0.57 - 2.65.

For most other materials this is not the case as most surfaces can be considered as dielectric. Even for metals, the surface will probably have oxidized.

Therefore the real refractive index interaction is strongly in favour of the electric vector and the complex part or imaginary Refractive Index can under these circumstances be simply considered as the absorption the light would encounter upon its passage through the particle.

Examples are:
- Polystyrene latex Refractive Index =1.62 - 0.0
- Alumina Refractive Index =1.56 - 0.1
- Glass beads Refractive Index =1.5 - 0.0

Light scattering occurs at the Refractive Index boundary between the media and the particle.

However with glass spheres Refractive Index ≈ 1.5 suspended in benzene Refractive Index ≈ 1.5 no measurable scattering is detected as the refractive indices are (virtually) identical.

A method of evaluation of the Refractive Index of a powder, involves suspending a well dispersed fraction in liquids of different Refractive Index until an index match is observed.

The Refractive Index of Isopropyl alcohol = 1.39. The Refractive Index of Methyl Naphthalene = 1.62. A mixture of equal volumes of each liquid well mixed together will in this case result in a binary liquid having a Refractive Index of 1.505. Mixtures of other ratios by volume enables a series of liquids to by prepared having any value of Refractive Index between 1.39 to 1.62.

4.3 Data Inversion

A particle size measurement results from a measurement of the scattering contributions made to each detector from each size of particle present.

In matrix notation we can write that:

$$\begin{pmatrix} L_1 \\ : \\ : \\ : \\ : \\ : \\ L_N \end{pmatrix} = \begin{pmatrix} T_N & & & & & \\ : & \searrow & & & & \\ : & & \searrow & & & \\ : & & & \searrow & & \\ : & & & & \searrow & \\ : & & & & & \searrow \\ : & & & & & T_{N,M} \end{pmatrix} \times \begin{pmatrix} S_1 \\ : \\ : \\ : \\ : \\ : \\ S_N \end{pmatrix}$$

$$L = TM \times S \qquad \text{Eq. 4-3}$$

In this form the light energy is seen to be the result of a matrix multiplication of size with the scattering matrix. However, the inverse of this equation is required.

$$S = TM^{-1} \times L \qquad \text{Eq. 4-4}$$

The equations are described as ill-posed and ill-conditioned making direct inversion without constraint un-viable. Therefore a degree of constraint is necessary which varies from manufacture dependent upon number of detectors, noise levels and experience.

Failure to adequately constrain the inversion leads to solutions of polydisperse distributions showing ripples in the histogram data. Serious lack of constraint can lead to zero or negative values and false modality.

The conclusion of the analysis is a particle size distribution providing the volume of particles that have a given size.

A value of the phase volume concentration is also provided as part of the normal result. This can and is used to confirm the choice of refractive index given to the particles by the method of challenging the machine with a suspension of particles having a known phase volume and confirming that the machine reports a similar value.

4.4 Calculation of Volumetric Concentrations and Specific Gravities

4.4.1 Calculation of Specific Gravities by Displacement

Place a measuring cylinder filled with 10 ml of solvent on a magnetic stirrer. Weigh the sample to be measured , then add in small portions to the solvent. When the volume of solution has reached a set point, e.g. 1 ml of solution has been displaced, weigh the sample again. The weight of well wetted and fully dispersed sample needed to displace 1 ml of solvent can then be calculated. For example the density of iso-octane is approximately 0.70, the weight of sample needed to displace one ml divided by 0.70 is the specific gravity.

4.4.2 Calculation of Volumetric Concentrations

It has been shown by Lips [1] that the true phase volume of a system of suspended spherical particles whose scattering extinction efficiencies have been correctly predicted by Mie theory can be measured correctly. It was shown that even for oblate spheroids of aspect ratio 5:1 the error was only of the order of 10%. Experiments confirmed that suspensions of known pre-determined phase volume was correctly reported by the Mastersizer. Therefore particles of a generally granular nature without excessive aspect ratio can be handled easily by this method which has been used to confirm that correct RI values have been applied to a given particle system. By combining the Mie theory of light scattering with the Beer-Lambert law the following equation is obtained.

$$c = \frac{100 \log_e (1 - \text{Obscuration})}{\frac{-3}{2} b \sum \frac{V_i Q_i}{d_i}}$$

Eq. 4-5

where c is the concentration (%), b is the beam length, V_i is the volume in sizeband I, Q_i is the extinction coefficient of sizeband I and d_i is the mean diameter of sizeband I. The extinction coefficient is a measure of how efficient a particle of a particular size is at scattering light.

Eq. 4-5 shows the relationship between volume concentration and the obscuration measured as part of a normal experiment. Once a correct particle size distribution has been established by measurement which assumes that suitable values for the RIs of the particle and the media have been used, then equation (1) can be used to provide a direct relationship between volume concentration and obscuration. This is providing the concentration of the dispersion is not too high (i.e. there is no multiple scattering).

4.5 Practical Results

The result shown in Fig. 4-8 was from a measurement of a Pigment powder that had been milled in a bead mill for some 38 hours. The optimum particle size is one that yields the highest liberation of colour. Over milling is very costly and rapidly reduces

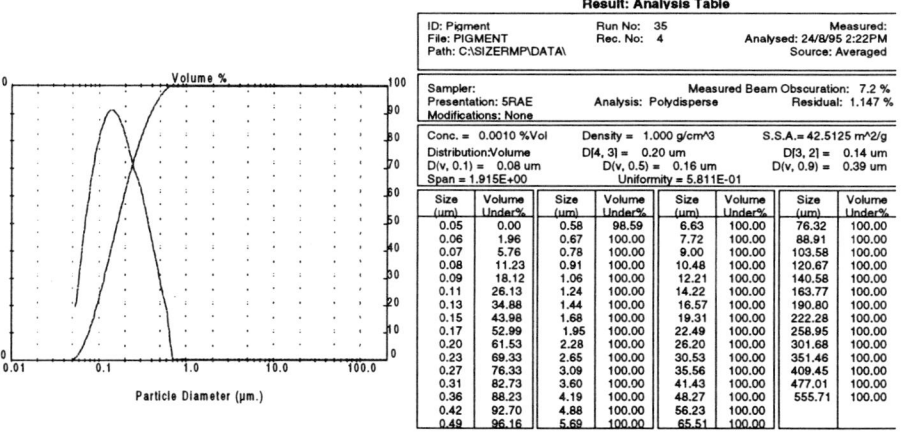

Fig. 4-8: Particle diameter of a pigment powder milled in a bead mill for some 38 hours

the colour yield if the size reduction process, is taken to far. Conversely undermilling requires that more expensive pigment be used to provide a given covering power. A further example is shown in Fig. 4-9. Intralipid is a vegetable emulsion in water. It is used for intraveneous feeding of patients unable to take nurishment by mouth. It is also required to remain as a stable dispersion even when drugs and or essential tissue salts.

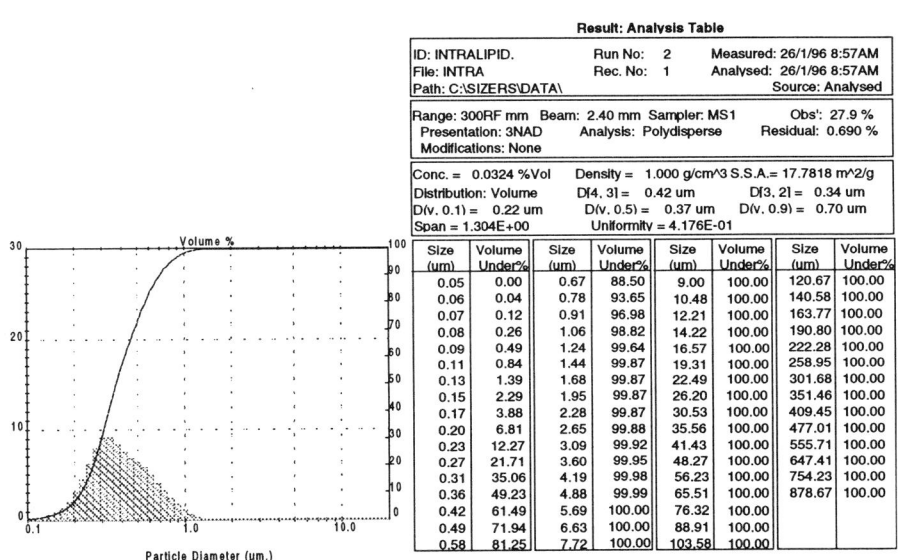

Fig. 4-9: Particle diameter of Intralipid®

4.6 References

(1) Lips, A. , Hart, P.M. and Evans, I.D., Proceedings of the 5th European Symposium in Particle Characterization, 1992, 443, Nurnberg Messe, Nurnberg.

(2) G. Mie, Ann, d. Physik (4), 25(1908), 377

(3) Principles of Optics, Eleletromagnetic Theory of Propogation, Interference aqnd Diffraction of Light, M. Born and E. Wolf, Pergamon Press

(4) Absorption and Scattering of Light by Small Particles, G.F. Boren and D.R. Huffman, Wiley - Interscience

(5) The Scattering of Light and other Electromagnetic Radiation, M. Kerker, Acedemic Press, 1969

manufacturing and other areas, to the International Organisation for Standardisation (ISO).

5.2.1.1 Good laboratory practice (GLP)

GLP evolved out of Good Manufacturing Practice (GMP) which was formally adopted by the FDA in 1979.

GLP regulations describe good laboratory practices that support applications for research or marketing licences for human and animal drugs and food products. GLP is formally defined in the UK as follows:

"Good Laboratory Practice (GLP) is concerned with the organisational processes and conditions under which the laboratory studies are planned, performed, monitored, recorded and reported.

Adherence by laboratories to the principles of GMP ensures the proper planning of studies and the provision of adequate means to carry them out. It facilitates the proper planning of studies, promotes their full and accurate reporting and provides a means whereby the integrity of the studies can be verified. The application of GLP to studies assures the quality and integrity of the data generated and allows its use by Government Regulatory Authorities in Hazard and Risk assessment of chemicals."

With regard to analytical instrument systems, GLP exists primarily to protect raw data. Raw data is normally defined as the first recorded paper or electronic representation of any item of data which is intended to support a Regulatory Submission. Protection is achieved by not allowing raw data to be altered and by recording any data manipulations, such as smoothing, in the form of an audit trail.

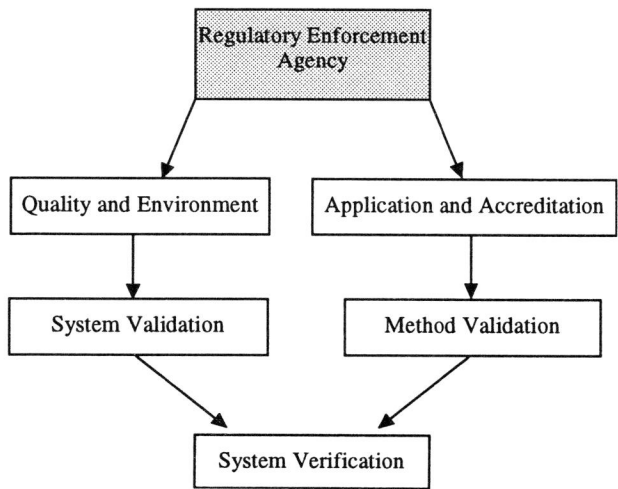

Fig. 5-1: The regulatory environment structure

How can the instrument supplier help?

By the provision of a "Qspec Validation Package"

Instrument **users** must provide clear and concise Standard Operating Procedures (SOPs). Apart from supplying instruments which are as easy to use as possible, **suppliers** can help by providing clear and correct operating instructions in a format that can be easily incorporated into the user's own SOP's.

As part of any Qspec Validation package, tailored training to provide first-time and mature users with suitable guidance in the formulation of robust and transferable Standard Operating Procedures which take full account of the properties of the material or compound being measured.

There are a number of other essential documentary items that the instrument supplier's software can provide including system identity ie. make, model and serial number; time and date of generation of data and manipulation history. These requirements are easily fulfilled if the supplier of the equipment has taken proper cognisance of the requirements of the Regulatory Authorities.

5.2.1.2 ISO

ISO 9000 is the international standard for a qualtiy management system first published in 1987. It consists of three levels of certification.

ISO 9001 is the most important standard since it encompasses both the design and the manufacturing requirements. This standard has been adopted in Europe as EN29000.

How can the instrument supplier help?

Laboratories which are GMP compliant may or may not seek ISO 9000 certification. However, they are increasingly recommended to seek compliance with this standard for any instruments supported. The standard does not of itself guarantee the quality of the instrument system but it does lay the groundwork and provide much of the essential documentation to facilitate validation.

5.2.2 Application and Accreditation

Accreditation is the procedure by which the capability of a laboratory to perform a specific range of tests or measurements is assessed. The accreditation covers the range of materials tested or analysed, the tests carried out, the method used and a variety of analytical parameters e.g. accuracy and precision. All these parameters are specific to the facility and the test.

Most national laboratory accreditation schemes are based on ISO/IEC Guide 25. In Europe this is provided as EN45001 in the UK, it is BS7501. The standard is implemented in the UK by the National Measurement and Accreditation Service (NAMAS), who contributed to a "memorandum of understanding", to the Western European Laboratory Accreditation Co-operation (WELAC), in 1990.

To date NAMAS has, or is in the process of obtaining, mutual recognition agreements with the following countries: Australia, Austria, Belgium, Denmark, Finland, France, Germany, Greece, Hong Kong, Iceland, Italy, Netherlands, New Zealand, Norway, Portugal, Spain and Switzerland.

Recently WELAC extended an invitation to the National Voluntary Laboratory Accreditation Program (NVLAP) in the USA to submit an application to WELAC for recognition as a member.

How can the instrument supplier help?

A NAMAS assessor proffered the following suggestions: "user friendly work instructions; easy self maintenance; advice on performance tests; use and provision of traceable calibration standards"

5.2.3 Method Validation

Method Validation ensures the integrity and quality of the analytical method.

The use of traceable primary and secondary sizing standards and specific routines for the verification of the performance of Sample Handling Units are key factors in meeting the requirements of Regulatory Authorities. These are the under followed in the draft ISO standard on laser diffraction.

How can the instrument supplier help?

The supplier can provide or propose a verification kit, comprising fully traceable size standards that will not only establish the correct performance of the optics and the software but will also confirm the ability of the sample handling system to maintain a properly proportioned volume distribution in each of the particle size classes.

The supplier can also provide data sets and system files to verify the correct performance of calculation routines and check the robustness of file handling and data transfer.

5.3 The Concept of Validation

5.3.1 Literature Definition.

Validation is a high level review to determine the conformity of a system, process or product to nationally established standards.

This can be assured by a "Quality Assurance" examination by an independent party which provides a high degree of confidence in the control of a system or process for managerial and regulatory purposes.

In the Pharmaceutical industry, validation is applied to Methods as well as Instruments and Software.

5.3.2 Definition of an Instrument

The modern definition of an instrument has been extended from a standalone analogue instrument such as a voltmeter to include complex microprocessor-based measurement devices controlled by microcomputers (such as laser diffraction particle sizes).

5.3.3 Generic System Validation

Fig. 5-1 shows a typical validation process for a regulatory environment. Regardless of whether a project concerns the erection of a new manufacturing facility or the provisioning of a new computer system, the protocol shown in Fig. 5-1 will still be applicable. The valid life cycle approach is also an accepted route and has similar objectives, this is shown in Fig. 5-2.

The various Qualification steps in the Q spec Protocol are:

5.3.3.1 Specification Qualification (SQ)/ Design Qualification (DQ)

This works on the premise that quality must be designed into a product from the outset. Regulatory bodies require evidence that:

- Rigorous design and specification methods are used.
  Full documentation the existence of Quality Control and Quality Assurance procedures.
- Suitably qualified and experienced personnel are used at all times.
  Comprehensive, planned testing is incorporated at all levels.
- Stringent change control, error reporting and corrective procedures are in place.

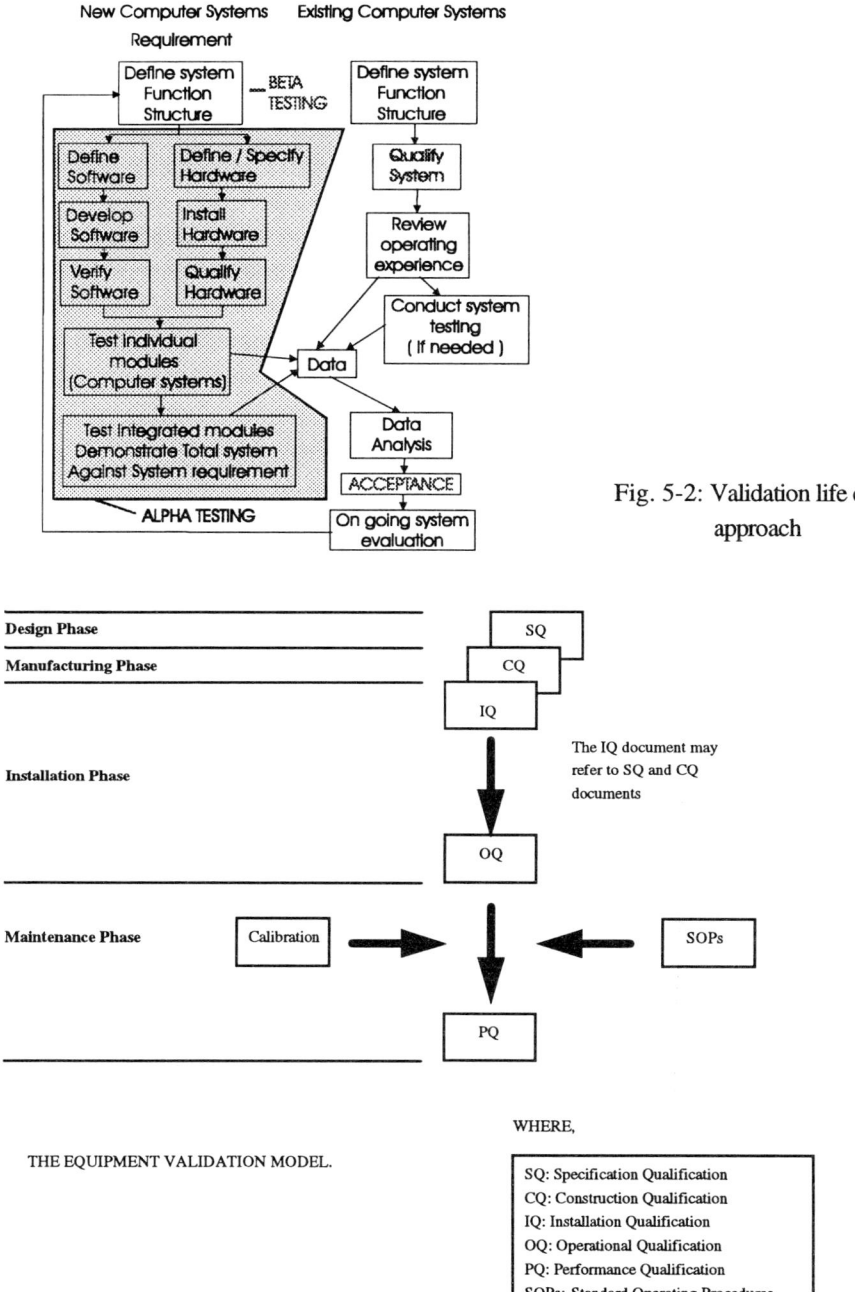

Fig. 5-2: Validation life cycle approach

Fig. 5-3: Overview of Validation Process (Q spec process)

The Concept of Validation 75

5.3.3.2 Installation Qualification (IQ)

Installation qualification involves the checking of equipment and control systems against the instrument suppliers standards. It also defines environment at physical safety and functional parameters prior to the initial utilisation of the system. It confirms that the system is correctly installed.

5.3.3.3 Operational qualification (OQ)

Operational qualification is the process of ensuring that the equipment performs consistently, as specified.
Applied to a particle size analyser, this process would establish the installed integrity of the system in the customer's environment.

5.3.3.4 Performance qualification (PQ)

Performance qualification is the process of testing normal operation when customers are manufacturing their own licenced products.
Unlike the other Qualification steps, Performance Qualification is the sole responsibility of the equipment user.
Having said this, suppliers can help by providing training which will enable the user to determine sensible and meaningful test methods relating to the specific process as well as giving guidance to the creation of Standard Operating Procedures.
The provision of such training can be negotiated as part of a Q-spec Validation contract.

Checklist for Regulatory Compliance of a LALLS (Low angle Laser Light Scattering) Particle Size Analyser.

Check
Data File
Does the Data File include the following information?
✓ Sample Identity
✓ Measurement parameters
✓ Time and date of measurement
✓ Unique system identity, such as a serial number
✓ Software version number
✓ Appropriate analysis algorithm and optical properties of the material
✓ Operator name

In addition:
Is there a way of confirming that a valid background measurement has preceded the measurement and that this is stored with the data file?
Can the operator store additional descriptive information with the data file?
Is it impossible for the the system to modify the raw data (after any input routines are completed) stored on the file?
Is the raw data preserved in the event of subsequent manipulation of the raw data?
Is there a visible record of data manipulations (a) on the file and (b) on the printed or displayed output?

Reports

Do reports clearly show the following information:
✓Sample identity
✓Method parameters
✓Time and date of measurement
✓Unique system identity, such as a serial number
✓Software version number
✓Appropriate analysis algorythm and assumed optical properties of the material
✓Operator name

Are all reports titled and page numbered?
Are filenames printed in reports?

System

Is the system delivered with a Certificate of System Verification?
Can the system be verified in the field?
Will the supplier of the system allow authorised regulatory authority representatives access to the software engineering documentation subject to non-disclosure agreement ?

5.3.4 Software Design and Validation

Software development follows the standard "V" life cycle (Fig. 5-4) which conforms to FDA guidelines, the IEEE Software Engineering Standards and TickIT.

5.3.5 Software environment

Software engineers should have adequate training as a minimum standard.
A secure software development replication environment must exist Software engineers should have password protective access to the Source Code Repository.

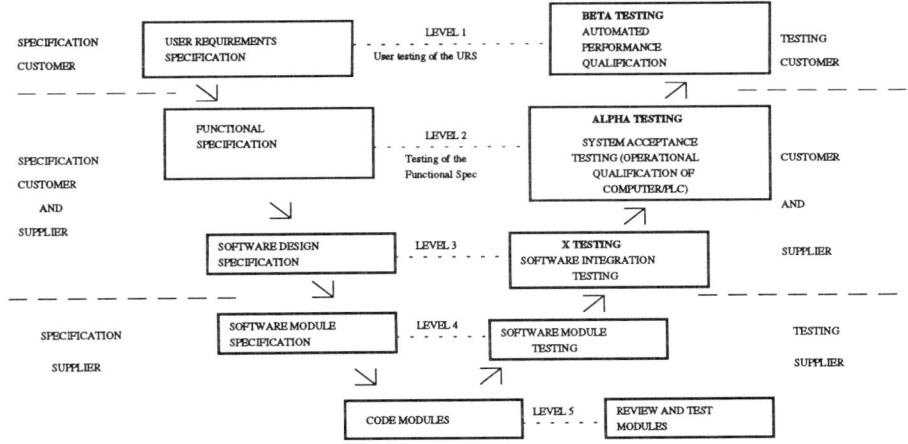

Fig. 5-4: Software Life Cycle Diagram

Software engineers should operate a configuration management system to control all manipulation of source code files.

The source code repository should backed up regularly with copies being held in a fire safe. Further copies should be archived off site. All computers used and all computer media used should be rigorously virus checked on a regular basis.

5.3.6 Change control

A Change Control system must be put in place whereby Software Change Requests are made with which detail and trace the requirements, authorisation and acceptance of each change. Through this medium, the Developer and any external auditors can verify that only authorised changes are made. Only authorised developers have access to the source modules. Procedures provide for the traceability of the authorisation as well as the change itself.

A formal right to view the software flow charts and life cycle documentation under security and or escrow agreements need to be in place.

A policy for any software bugs which has safety implications or causes the system to report false or misleading data or which causes the software to crash must be established. This could include:

- Shipment of software is replaced by the previous version complete with documentation, obtaining the dangers..
- Fixes for less serious, irritating and non fatal bugs must be dealt with in future releases of the software.
- A notification procedure whereby the end user can announce the finding of any software bugs.

Every version of released software, together with the documentation is archived by the supplier and at least two sites in accordance with FDA recommendations.

5.3.7 Verification of LALLS Particle Size analysers

The most obvious question facing anyone considering the verification of a LALLS Particle size analyser is " Does it give the right answer". A second question which will be posed by someone unfamiliar with the technique will often be "How is the instrument calibrated?" The short answer to this question is that not only is the instrument not calibrated but it can't be calibrated within the following strict definition:

"Calibration is a process where an instrument is used to measure a known standard and its response is adjusted until the answer given corresponds to the standard. The instrument will then hold this adjustment for a limited period of time and will then gradually depart from the calibrated state."

According to this definition, it is not possible to "calibrate" a LALLS instrument since the measurement is based on First Principles and the answers that it gives can not be adjusted.

Obviously, the concern about calibration arises from a need to know whether the instrument is giving a correct answer and this need is met by a process known as "verification". This is a process of testing the instrument's responses to a set of known particle size standards in order to check that they fall within acceptable tolerances.

At the time of publication of this document, an International Standards Organisation (ISO) standard for LALLS particle size analysers is in draft.

This standard recognises the need for the use of spherical materials since, in common with all sizing techniques, LALLS assumes that particles are spherical. (In the past, other standards available from the European Community Bureau of Reference (BCR) have been used but since these are quartz materials which are irregularly shaped, they remain less suitable for the verification of LALLS instruments within these regulated envioroments as the results will be different dependent upon the measurement technique employed)

5.3.8 Primary Verification

Two verification routes are proposed. The first, or primary route is to use the spherical Certified Reference Material (CRM), of which current choices are restricted to NBS 1003b and NBS 1004a glass beads. These are available from the US National Institute of Science and Technology (NIST).

5.3.9 Secondary Verification

Monosized latex beads are not ideal since, although they can prove the instrument's ability to correctly identify a size, they do not challenge the more important capability of an instrument to correctly describe a broad distribution of sizes and to verify that the correct volume of material has been subscribed to each size.

The number of available spherical CRMs is due to be increased with the production of an extensive range of materials in decades of sizes from 0.1 up to 650 microns. However, these are not expected to be released before 1997.

An important consideration in the use of primary CRMs is that they are not necessarily a reliable way of stand-alone testing of the optical setup, electronics and software of a LALLS instrument since the accuracy of the results obtained will not only be prone to errors in sampling techniques but will be influenced by variations in sample dispersion techniques. Differences in the design of sample presentation units can also be a significant source of variations in measurements. Having said this, it is obvious that CRMs are essential for whole-body testing of LALLS instruments.

Within the optical system certain single dimensional settings will require verification. These may be achieved with monosizer latex. However, latex calibrats are not generally suitable.

The need for a standard which can be used to check the optical setup and the detector of an instrument in isolation while being uninfluenced by variations in storage or sample handling of real samples has given rise to the production of a secondary validation device known as a Reference reticle. This is a an ensemble of chrome dots of known size certified to the primary standard of length which have been deposited onto an optical glass substrate. This is placed in the laser beam of the instrument at the sampling plane and the resulting scattering pattern is then recorded. Analysis of this pattern will result in the production of a particle size distribution which can be compared with the known certified distribution from the reticle.

The use of a traceable reference reticle has been a fundamental part of Malvern Instruments procedures since the mid 1980's.

5.3.10 Routine Verification

As CRMs are limited in number and expensive, they are unlikely to be used for daily or weekly verification of LALLS instruments.

This is best carried out using a generally available material which is robust, chemically inert and as near spherical as possible.

Typical of such materials are the solid glass beads used in the surface treatment industry for air blasting. Such material can be obtained from:

Abrasive Developments Ltd, Norman House, Henley-in-Arden, Solihull, West Midlands B95 5AH (01564 - 792231)

Several grades of material are available and the grade chosen should correspond in size to the size range of the material being measured daily. One drawback to the use of glass beads is that they are prone to stratification in transit or handling and need to be very thoroughly mixed to ensure correct sampling. Our experience has shown that the best method of sampling is to fill the sample container two thirds full, close the container and tumble it hand-over-hand for at least five minutes before extracting a sample for measurement. When this is done, the full reproducibility of the instrument can be seen to full effect.

Once a suitable material has been found, it should be measured and characterised immediately after a successful verification of the instrument has been completed. The results obtained can then be preserved as an internal reference document against which subsequent analyses can be compared.

A daily or weekly verification (calibration) record can be made to ensure that no drift or departure from normal has taken place.

5.3.11 Software Validation

The main problem with software validation is that software can **not** be retrospectively validated. Unless the software has been developed in accordance with the protocols outlined earlier in this Handbook, there is no means of determining that the software is absolutely free from problems which may arise from code which is redundant or poorly structured or from poor error trapping. Subscribers to any QSpec Contract will have rights to view the life cycle documentation and flow charts as well as having individual Escrow contracts giving them rights to access to the source code in the event of failure of the suppliers company.

Users of software which predates these proceedures can meet the requirements of the Regulatory bodies on a case-by-case or concessionary basis by carrying out certain tests to establish the numeric accuracy of the measurement routines embodied in the software.

5.3.12 Numerical and Analytical Verification of non-Validated software

The fundamental data gathered by a LALLS instrument is the light energy distribution from the elements of the detector. The acid test of the analysis routines embodied in the software is to verify that this data is reliably converted into an accurate particle size distribution. This is done to a large extent by the use of primary and secondary standards as described earlier.

Another important aspect of the analysis which also needs to be verified is the correct calculation of the derived diameters. These diameters which are used extensively in product specifications and submissions are the various moments of the particle size distribution such as the volume percentiles ($D_{v10}; D_{v50}; D_{v90}$ etc, the volume mean ($D_{4,3}$), the surface mean ($D_{3,2}$) etc. Verification of the numeracy of LALLS instruments can be carried out by generating a normal particle size distribution and manually calculating these moments.

5.3.13 Analytical Light Energy Distributions

LALLS instruments depend on the principle of predicting the light scattering behaviour of various sized particles when they are illuminated by a single wavelength light source.

The light scattering pattern produced by an ensemble of particles is observed and recorded by a multi-element detector. Individual elements of this detector provide the software with electrical currents proportional to the intensity of the scattered light contributed by particles within the sample. From light scattering theory (Fraunhofer, Anomalous diffraction or Mie theory) the contribution of light scattered from each size of particle to all the detector elements can be constructed and a matrix of data stored in the computer memory.

In some implementations of the measurement, we observe the light scattered by an unknown system of particles as observed values of current as described above. A "starting estimate" of the sample particle size distribution is created within the software and the light scattering pattern which would be produced by this trial particle size distribution is predicted using the relevant light scattering theory. This predicted pattern is then compared with the actual pattern observed. The initial trial particle size distribution is then modified and the scattering pattern which this would produce is compared with the measured pattern. This process is continued iteratively using a fitting routine until the predicted scattering pattern is in close agreement with the measured pattern. At this point the particle size distribution which produces the matching pattern is output as the result of the analysis together with a quality value for the fit.

As part of a development toolkit, one is able to calculate the theoretical light energy distributions for any given particle size distribution and this facility has enabled standard test data to be produced to test the numeracy of the inversion routine.

This data can be supplied either in a magnetic form for direct input into the software to be tested or in tabular form for keyboard entry. Once this data has been input, it is possible to verify the correctness of the inversion routines built into the software under audit.

Owing to the fact that the ultimate responsibility for validation rests with the user, this data and the instructions detailing its correct use are supplied to the user as a Verification "kit". This kit also includes standard result files together with file saving and manipulation routines which enable the user to verify the integrity of the file handling routines in the software being tested.

5.3.14 Software Verification

Modern measurement instruments deliver a vast array of calculated results from the data they obtain.

Most software is supplied in the form of compiled code. Thus the full operation can only be appreciated by either undertaking the arduous and skilled task of reading and checking the source code and or by comparing one set of software with another, whose language and structure is entirely different.

A spreadsheet package such as Excel (TM) are not required to be validated as such. Their vast usage within industry in general, provides an on-going continuous peer review. Faults and bugs are rapidly communicated and are freely available in the public domain.

It is therefore possible to verify a closed set of capital software with an open and published set if an Excel spreadsheet is employed.

Such a route is shown in Fig. 5-5. The numerical integrity of the closed software package is assured by conducting any defined function or calculation both within the closed package and within the open Excel package and comparing the

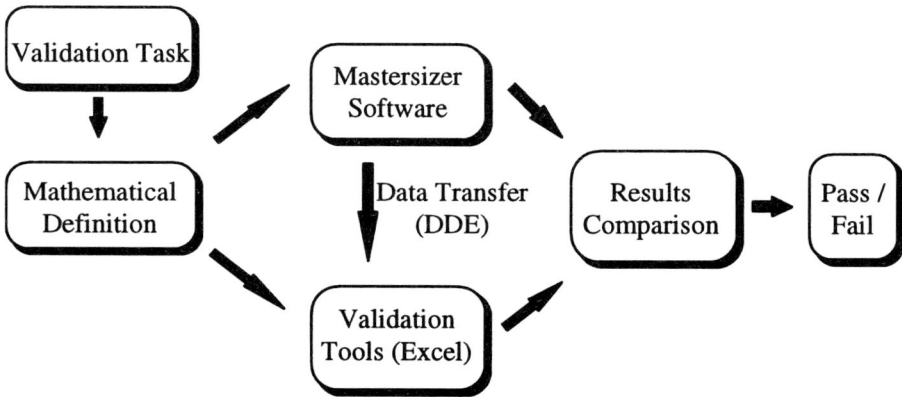

Fig. 5-5: Verification of a closed set of capital software with an open and published set.

result. A typical example of a series of calculation is shown in Eq. 5-1 and Tab. 5-1. A comprehensive report dealing with each result which is the subject of a data manipulation or calculation can be made available in support of the numeracy and functionality of the software version under test.

$$D[M,N] = \left[\frac{\sum_{i=1}^{m} p_v(i)d(i)^{M-3}}{\sum_{i=1}^{m} p_v(i)d(i)^{N-3}} \right]^{\frac{1}{M-N}} \quad M = 1,2,3,4; \ N = 0,1,2,3; \ M > N$$

Eq. 5-1

Where:
- $d(i)$ Mean diameter in size band $i = 1,2,...,m$
- $p_v(i)$ Volume distribution $i = 1,2,...,m$
- m Number of result channels

Tab. 5-1: Validation result moments calculation

Validation Result Moments Calculation					
	Sizer Output	Validation	Error	Tolerance	Pass/Fail
D (4,3)	40.71	40.71	-1.77E-05	+/-2.00%	Pass
D (4,2)	40.00	40.00	-2.97E-06	+/-2.00%	Pass
D (4,1)	39.30	39.30	2.63E-05	+/-2.00%	Pass
D (4,0)	38.62	38.62	-2.68E-05	+/-2.00%	Pass
D (3,2)	39.30	39.30	3.20E-05	+/-2.00%	Pass
D (3,1)	38.62	38.62	-4.02E-05	+/-2.00%	Pass
D (3,0)	37.94	37.94	-4.77E-05	+/-2.00%	Pass
D (2,1)	37.94	37.94	3.74E-05	+/-2.00%	Pass
D (2,0)	37.28	37.28	4.77E-05	+/-2.00%	Pass
D (1,0)	36.63	36.63	2.53E-05	+/-2.00%	Pass

Address of the Author:
Maurice Wedd
Malvern Instruments Limited
Spring Lane South, Malvern
Worcestershire WR14 1AT
United Kingdom

6 Confocal Laser Scanning Microscopy and its Application in Liposomal Research

T. Möller, Max-Delbrück-Centrum, Berlin

6.1 Summary

Confocal laser scanning microscopy is a modern optical microscopic technique which offers significant advantages over conventional microscopy. It reduces the fluorescence blur from out-of-focus structures and enhances image resolution both laterally and axially. Confocal microscopy allows to optically section specimens which prevents physical sectioning artifacts observed with light and electron microscopy. Because optical sectioning is non-invasive, living, as well as, fixed cells can be observed with greater clarity. These advantages are achieved by an advanced optical design. The specimen is illuminated by a diffraction limited spot of laserlight. Fluorescence excited in the volume element (voxel) is focused on an aperture confocal with the in-focus voxel. The light passing through the aperture is registered by a photodetector. While the spot is scanned over the specimen, the electrical output from the detector is displayed on a computerscreen, thus building up a two-dimensional image with factor ~ $2^{1/2}$ increased resolution. Serial sections allow, with the help of computers, 3D reconstructions of the scanned specimen which can give new insights into the object under investigation.

6.2 Introduction

Light microscopy is an important tool in cell biology. It is used to analyze cell structure, physiology and function in living, as well as fixed cells, and tissues. Conventional epi-fluorescence microscopy is widely used for immunocytochemical localization of cellular antigens. Both techniques suffer from this disadvantage: Out-of-focus structures often obscure structures of interest. In the case of the epi-fluorescence, this means that illumination of the entire field of view with light at the excitatory wavelength, excites fluorescence throughout the whole depth of the specimen, rather than just at the focal plane (Fig. 6-1A). Much of the fluorescence coming from regions above and below the specimen is also collected by the objective and thus accounts to the out-of-focus blur of the image at this particular focal plane.

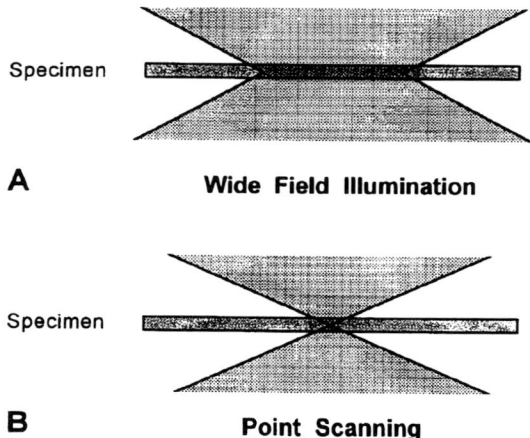

Fig. 6-1: Comparison of specimen illumination in conventional and confocal microscopes. In conventional microscopy, as shown in A, the whole depth of the specimen is illuminated at the same time. This causes signals not only in the plane of interest, but also out-of-focus blur. In contrast, a confocal microscope (B) illuminates only a portion of the specimen at one time. This strongly reduces the out-of-focus blur. By scanning the spot over the entire specimen, a complete image of the specimen in the plane of focus is generated.

This blurring seriously degrades the quality of the image by reducing its sharpness and contrast.
Confocal laser scanning microscopy (CLSM) is a modern optical technique that overcomes this problem, adding several advantages over conventional epi-fluorescence microscopy. Beside the virtually out-of-focus blur free image, it improves the lateral, as well as, the axial resolution and gives the capability of non-invasive serial optical sectioning of intact, thick specimens. These serial sections can be used to calculate the three-dimensional (3D) distribution of labels within cells or tissues.

6.3 Principles of Confocal Microscopy

As stated above, conventional microscopy causes the problem of out-of focus blur by the uniform and simultaneous illumination of the entire specimen. Confocal microscopy illuminates sequentialy, not simultaneously. The light is focused as a spot on one volume element (voxel) of the specimen at a time (Fig. 6-1B). The bright waist of the hourglass-shaped beam strikes only one spot of the specimen in a chosen depth.

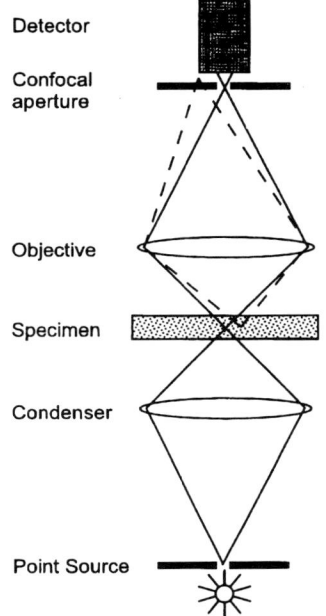

Fig. 6-2: Principle of a confocal microscope. The light of a point source is focused by the condenser onto the specimen. Light passing through the specimen or fluorescence excited in the specimen is collected by the lens of the objective and focused onto the detector pinhole. In-focus light passes through the spatial filter, whereas out-of-focus light from other focal planes is blocked. So, mainly in-focus light/ fluorescence is detected.

This reduces the illumination and excited fluorescence intensity above and below the focal plane. By scanning the spot over the entire specimen, a complete image of this focal plane is produced. The second step to remove the now reduced, but still existing out-of-focus blur is based on the principle that both the illumination and the detection systems are focused on the same single voxel of the specimen (Fig. 6-2). This removal is achieved by a spatial filter (pinhole), which is placed in the emitting light path in front of the detector (Fig. 6-2). As illumination, specimen and detector are all in the same focal plane, now they are called confocal.

Interestingly, this principle was invented by Marvin Minsky, better know as the father of artificial intelligence. In the 1950s, as a postdoctoral fellow at Harvard University, he built a revolutionary light microscope, which he patented in 1957 as a „double-focusing stage-scanning microscope". In the patent application he pointed out several advantages:

- Reduced blurring of the image from light scattering
- Increased resolution
- Improved signal-to-noise ratio
- Unusually clear examination of thick light-scattering objects
- xy-scan possible over wide areas of the specimen
- Inclusion of a z-scan is possible
- Electronic adjustment of magnification

In the 17-year life of the patent, no one was interested in an instrument of this design. It took about 25 years until the principle of confocal microscopy had a splendid rebirth.

Modern confocal microscopes, commercially available since the late '80s, use a laser as the light source (Fig. 6-3). The beam of light is deflected by dichroic mirror and passes through scanning system and reaches the objective, which focuses the scanning beam as a spot on the specimen. The excited fluorescence in the specimen returns via the objective and the scanning system to the dichroic mirror. Because the fluorescence has a longer wavelength (loose in energy) than the exciting laser light, it is not deflected by the dichroic mirror and so focused on the confocal aperture in front of the detector. Most fluorescence originating from above or below the plane of focus in the specimen does not pass through the aperture (pinhole diaphragm) and, as a result, only a little of the out-of-focus fluorescence reaches the detector. This is the „major secret" of confocal microscopy. Elimination of out-of-focus blur by spatial filtering using a point source of light for excitation and a confocal pinhole in front of the detector.

6.4 Resolution

Without going to much into detail, let us recall some basics about optical resolution. About a century ago, Ernst Abbé established the foundation of light microscopy. He showed that the diffraction of light by the specimen and by the objective lens determined image resolution and established the role of the objective and condenser numerical apertures (Eq. 6-1).

$$d_{min} = 1.22 \times \frac{\lambda_o}{NA_{obj} + NA_{cond}} \qquad \text{Eq. 6-1}$$

where d_{min} is the minimum distance in a periodic grating that the diffraction images of two points in the specimen can approach each other laterally before they merge and can no longer be resolved as two separate points. d is expressed as distance within the focal plane in the specimen space. λ_0 is the wavelength of the light in vacuum and NA_{obj} and NA_{cond} are the numerical apertures of the objective and the condenser respectively. The numerical aperture is the product of the sine of the half-angle (α) of the cone of light either acceptable by the objective lens or emerging from the condenser lens and the refractive indexes (η) of the medium between the specimen and the objective or condenser lens, respectively. This equation clearly shows, that the achieved lateral resolution at a given wavelength increases with higher numerical apertures for the objective and condenser, respectively. In epi-fluorescence, where the objective also serves as the condenser, the resolution depends only on the numerical

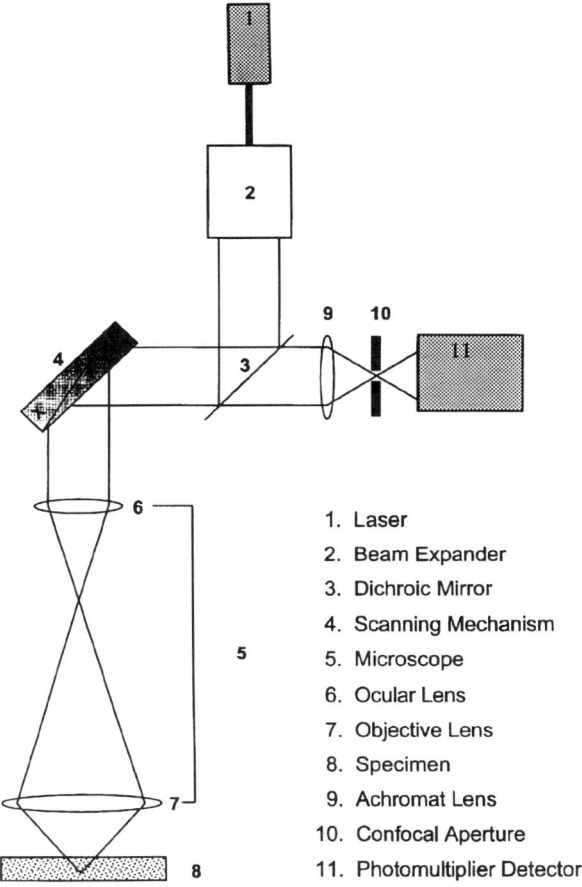

Fig. 6-3: Design of a contemporary confocal microscope. A laser is used as the point source of light. A beam expander (a line of lenses) widens the beam of light. It is deflected by dichroic mirror and passes through scanning system and reaches the objective, which focuses the scanning beam as a spot on the specimen. The excited fluorescence in the specimen returns via the objective and the scanning system to the dichroic mirror. Because the fluorescence got a longer wavelength (loose in energy) than the exciting laser light, it is not deflected by the dichroic mirror and so focused on the confocal aperture in front of the detector. Most fluorescence originating from above or below the plane of focus in the specimen does not pass through the aperture (pinhole diaphragm) and, as a result, only little of the out-of-focus fluorescence reaches the detector.

aperture of the objective. It is also immediately understandable that reducing the wavelength of the light is increasing the resolution.

The theory gets more complicated when we consider axial resolution and starts to get impossible when we think about resolution in confocal microscopy, we will not deal with any equations, but concentrate on the functional implications. As in normal epifluorescence microscopy, the resolution in confocal microscopy depends, at a given wavelength, mainly on the numerical aperture. Tab. 6-1 shows some examples of theoretical resolutions in widefield and confocal microscopy.

Tab. 6-1: Theoretical microscope resolutions. The Table shows the dependence of the theoretical resolution on the numerical aperture (NA) and the wavelength. For confocal fluorescence microscopy the wavelengths are given as the excitation/emission pair. Objectives are stated by their magnification and numerical aperture. The given resolutions for conventional microscopy are calculated according to standard equations given in Handbook of Biological Confocal Microscopy 2nd Ed., „Foundations of confocal scanned imaging in light microscopy", S. Inoué and for confocal fluorescence microscopy according to the equations given in [1].

	Lateral resolution (µm)		Axial resolution (µm)	
Wavelength (nm)	x40 NA 0.85	x60 NA 1.40	x40 NA 0.85	x60 NA 1.40
	Conventional microscopy			
351	0,25	0,17	0,97	0,62
488	0,35	0,23	1,36	0,89
514	0,37	0,24	1,43	0,92
568	0,41	0,27	1,56	1,02
	Confocal fluorescence microscopy			
351/450	0,20	0,12	0,77	0,46
488/518	0,25	0,16	0,98	0,56
514/540	0,27	0,17	1,03	0,58
568/590	0,30	0,18	1,09	0,64

A second important influence on resolution is the size of the out-of-focus blur rejecting pinhole. All the considerable advantages of confocal microscopy derive from the small size of this aperture in front of the detector. A small confocal aperture is essential if the maximum resolution and optical sectioning resolution are to be realized concurrently. A larger pinhole may be required to increase the signal when there is only a limited or fading fluorescence. The actual size of the pinhole depends on the designs of the confocal microscope and they are somewhat incomparable. Yet, the lateral resolution is more sensitive to the pinhole size than the axial resolution.

Another factor which influences resolution is the wavelength of light. Tab. 6-1 clearly shows that using shorter wavelengths increases resolution. As the wavelength depend on the lightsource, we will discuss lasers, the light source in most contemporary confocal microscopes, in the next section.

Overall, for fluorescent imaging the resolution confocal microscopy can be a factor of $~2^{1/2}$ greater than achieved with conventional microscopy.

6.5 Lasers

Lasers are the commonly used light sources in confocal microscopy. There is a large variety of technologies, we will not discuss here in detail. They all have a number of unique properties, which makes them an almost ideal light source for use in confocal microscopy:
- High degree of monochromaticity
- Small divergence
- High brightness
- High degree of spatial and temporal coherence

Beside these advantages, there is one major disadvantage. A laser usually emits only one or very few wavelengths. Although there are many types of laser available the number of wavelengths is still limited. The following table (Tab. 6-2) shows some lasers often used in confocal microscopy.

As a consequence, the limitations in available (payable) laser wavelengths also restrict the number of useable dyes in confocal laser scanning microscopy.

Tab. 6-2: Lasers often used in confocal microscopy

Wavelength (nm)	Power (mW)	Type of laser
351	300	water-cooled argon UV laser
364	>25	air-cooled argon UV laser
457/488/514	25	small-frame argon-ion laser
543	1,2	green He-Ne laser
633	3,5	He-Ne laser
1152	1	He-Ne laser

6.6 Dyes

There are two major forms of dyes which are used in confocal microscopy. Fluorophores that can be tagged onto structures of interest to show their location within the specimen and dyes with optical properties that are sensitive to changes in the local environment. They all have one feature in common. Their excitation wavelength has to correspond to a laser wavelength. In fact, some well established fluorophores have been modified to meet the needs of confocal laser scanning microscopy.

The dyes for localization are either site-specific, or they can be tagged to site-specific macromolecules (e.g. antibodies). The fluorophores which detect changes in their environment are used in physiology to measure, for example, calcium concentrations, pH and membrane potential. Tab. 6-3 shows some common dyes, their spectral properties and application.

6.7 Limits of Confocal Microscopy

Besides the valuable advantages described in the previous sections, there are some limitations of confocal microscopy. In addition to the restriction of available wavelengths, there are photon efficiency of the intermediate optical system, detection and measurement losses as well as possible specimen damage, which have to be considered. The photon efficiency of the intermediate optical system is strongly influenced by the quality of the lenses and mirrors. Even as the contemporary commercial confocal microscopes all use high quality optical parts, we still have to consider, that additional elements in the optical pathway cause additional losses. In the case of the detection systems, the photomultipliers have only a limited quantum efficiency (i.e. the portion of photons arriving at the detector that actually contribute to its output signal) and an inherent noise level. The quantum efficiency of a photomultiplier may be has high as 30% for blue and green light, but this still means that 70% of the photons produce no signal. In the case of red light the efficiency is even lower. All the possible reductions in fluorescence intensity may cause a change for the worse in the signal-to-noise ratio, producing detection problems at very weak fluorescence levels. As we use a highly focused spot of laser light to scan the specimen, one can imagine that this may not only cause photobleaching, but at high laser power, we also "cook" the specimen.

Tab. 6-3: The table shows the application and spectral properties of some fluorescent dyes. The values are given for appropriate measurement conditions as laid down by their distributors.

Application	Fluorophore	Absorption max. (nm)	Emission max. (nm)
Covalent labeling reagents	Fluorescein-isothiocyanat (FITC)	490	520
	Tetramethylrodamine-isothiocyanat (TRITC)	554	573
	Texas Red®	596	620
	BODIPY series	500 - 581	510 - 591
	CY3	554	568
	CY5	652	672
	CY7	755	778
DNA content	Hoechst	340	450
	DAPI	350	470
	Ethidium bromide	510	595
Expression label	Green fluorescent protein	395/470	509
Membrane location and fluidity	DiO	484	501
	DiA	491	613
	DiI-C18-(3)	546	565
	Texas Red®-phospad-idylethanolamine	590	620
Calcium	Fura-2	335 (low Ca^{2+})	515 (low Ca^{2+})
		360 (high Ca^{2+})	510 (high Ca^{2+})
	Indo-1	330 (high Ca^{2+})	400 (high Ca^{2+})
		350 (low Ca^{2+})	485 (low Ca^{2+})
	Fluo-3	506	526
pH	BCECF	460 (low pH)	530 (low pH)
		505 (high pH)	530 (high pH)
	SNARF-1	518/548 (low pH)	587 (low pH)
		574 (high pH)	636 (high pH)
Membrane potential	DiO-Cn-(3)	485	505
	Rhodamine123	511	534

Nevertheless, confocal microscopy is a very sensitive, and in case of resolution, not matched fluorescence detection technique. We just have to keep in mind, that it is not a cure-all and it has some inherent limitations.

6.8 Applications of Confocal Microscopy

Confocal microscopy is a powerful tool to investigate cellular structure and physiology. As discussed in the Dyes section, we have fluorophores which are sensitive to changes in their microenvironment (i.e. calcium concentrations, pH, etc.). Another application for confocal microscopy is the localization of structures within a cell or tissue. This includes cellular localization of organelles, cytoskeletal elements and macromolecules as proteins, DNA and RNA, as well as studying the interaction of cells with colloidal particles, for example, liposomes. Especially useful is the imaging of multiple labels, where two or more structures can be observed in the same cell or tissue, allowing us to trace the fate of colloidal particles. For example, the particle is labeled with one color (e.g. red), whereas the cell, or cell compartment of interest carries another color (e.g. green). The possibilities of labeling liposomes and their target cells will be discussed in the next section.

6.8.1 Detection of Immunoliposome-Cell Interaction

Among the colloidal particulate systems proposed for site-specific drug delivery, liposomes have attracted considerable attention. Liposomes consist of one or more concentric phospholipid bilayers separated by one or more internal aqueous compartments. A large variety of drugs (e.g. anti-tumor and anti-microbial drugs, enzymes, peptides, hormones, and genetic material) have already been encapsulated in either the aqueous or the lipid phase of liposomes. To enhance target site binding, liposomes can be conjugated with various homing devices. Monoclonal or polyclonal antibodies have proven to be the most valuable homing devices for liposomes.

To detect immunoliposomes with confocal laser scanning microscopy, they have to be labeled by a fluorophore (this is true for all systems under study). There are three possibilities to do so. First, labeling of the lipid phase; second, labeling of the aqueous phase; and third, labeling of the carried substance.

6.8.1.1 Labeling of the Lipid Phase

The most commonly used fluorophores to label membranes are DiI, DiO, DiA and Texas Red ® PE. Tab. 6-3 shows their spectral properties. There are in fact more dyes available in the catalogues of fluorescent probe providers, which differ in membrane integration properties and spectral properties. So for almost every need, there should be a suitable dye. As labeling of the liposomal membrane means integration of an additional substance in the lipid bilayer, it might cause some interferences in the attachment of targeting devices as well as sterical problems in cell-particle interaction.

6.8.1.2 Labeling the Aqueous Phase

In principle, all watersoluble fluorophores can be used. As some dyes are cytotoxic, they should not be used if the fate of internalized liposomes is studied. One dye, Carboxy fluorescein (CF), is widely used to label the aqueous compartment of liposomes, for fusion studies of the liposomes with cells. Very high concentrations of CF reduce its own fluorescence signal by a process called quenching. Loading liposomes with CF at quenching concentrations results in vesicles with very high concentrated fluorophore but a limited fluorescence. If liposomes fuse with their target cells, the CF is released and diluted to a concentration where fluorescence is not quenched. Even if only few numbers of liposomes fuse with a cell this „brighter if unpacked" dye produces enough fluorescence to be detected.

6.8.1.3 Labeling of carried substances

Depending on the substance which is carried (e.g. drugs, DNA, proteins) different labeling protocols exist. Labeling the carried substance gives the possibility to study their fate within the targeted cell, but also limits the amount of fluorophore which can be carried.

6.8.1.4 Labeling of Cells and Cell Compartments

As the cell or cell compartments are the targets for the colloidal particulate systems, they must be labeled for study. This labeling should not interfere with the label of the liposomes and due to the broad variety of dyes, this is not a problem.

Cell membranes can also be labeled by the same fluorophores as mentioned above. There are dyes which, for example, stain site-specific ER membranes or mitochondria. Another more specific method is to stain surface antigens by specific antibodies. This, in fact, can be done, with any epitope in and outside the cell, where specific antibodies are available. For example, an antibody against an endosomal antigen can be used to study possible internalization of the vesicles. This double-labeling gives a very accurate method to look for site-specific interactions.

Fig. 6-4 and Fig. 6-5 show some examples of liposome-cell interaction.

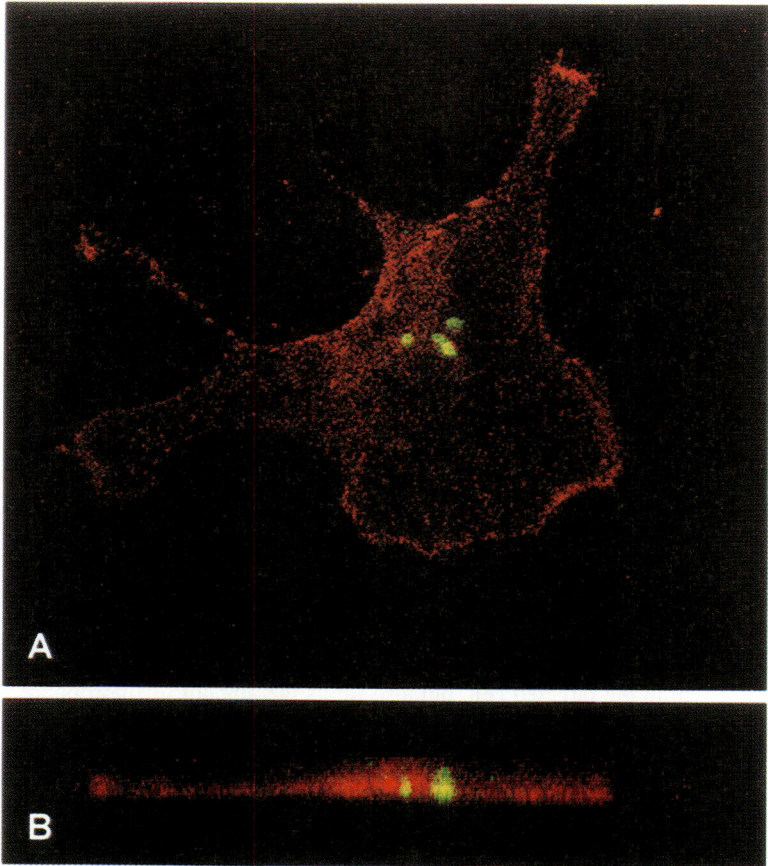

Fig. 6-4: Localization of liposomes in rat mesengial cells. The red/green image shows the Carboxy fluorescein (CF) loaded liposomes in the green and the Cy3 stained cell surface superimposed in the red channel. Virtual CLSM cross sections of the cells showed, that the CF was located intracellular (A+B). The dye was stored in vesicular structures. In contrast, control anchor liposomes (prepared in the same way but without antibody) failed to show any CF in or on the cells (not shown). Primary cultures of rat mesengial cells (RMC) were seeded onto glass slides, grown until they reached 60% confluence, washed twice with PBS and incubated with 100 µl immunoliposomes containing the fluorescence dye carboxy fluorescein (CF). The suspension was dissolved in a total volume of 2 ml media per slide and incubated at 37°C for 5 hours to elucidate the subsequent fate of liposomes. The surface of the RMC was stained after fixation with mouse anti-Thy 1.1. antibodies, which where detected by a CY3-coupled commercial anti-mouse antibody (red). [2]

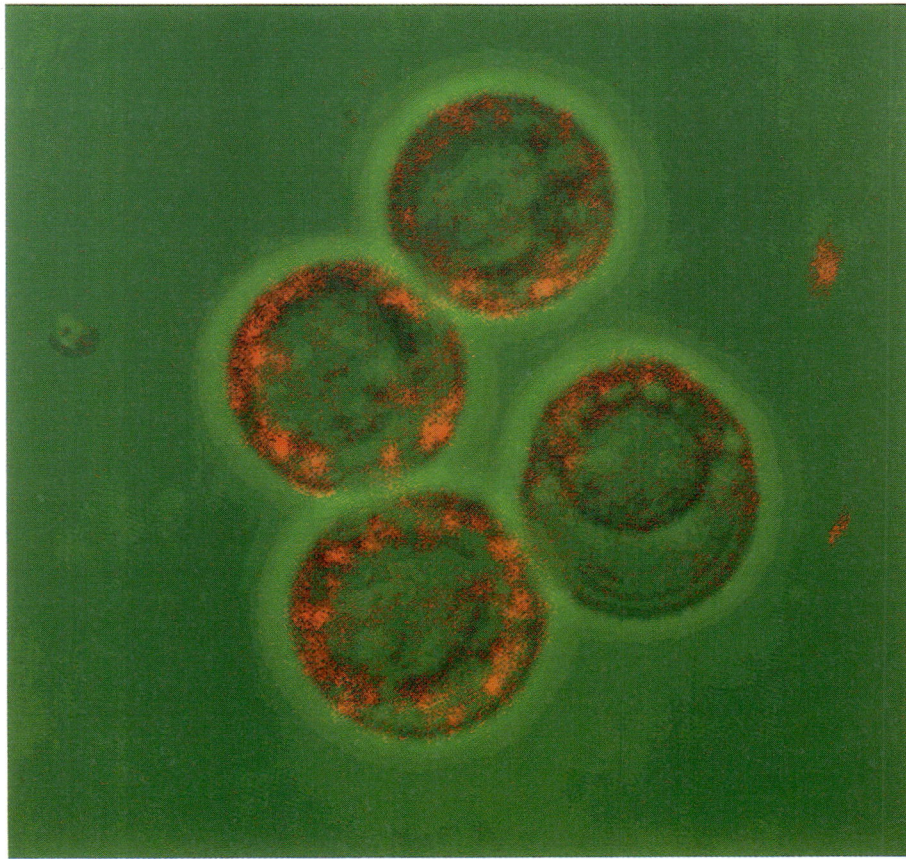

Fig. 6-5: Internalization of anti-ICAM-1 immunoliposomes in human bronchial epithelial cells (BEAS-2B). The red/green image shows the Texas-Red®-PE labeled liposomes in the red channel superimposed by the cell outline of a phasecontrast image in the green channel. The liposomes are located within the cell. The anti-ICAM-1 monoclonal antibody F10.2 was conjugated to liposomes to target cells expressing the cell adhesion molecule ICAM-1. The lipid composition of the liposomes used was partially hydrogenated egg-phosphatidylcholine (PC), cholesterol (CH) and N-[4-(p-maleimidophenyl)butyryl] phosphatidylethanolamine (MPB-PE] and Texas Red® -phosphatidylethanolamine (PE) at a molar ratio of 10/1.4/0.4/0.01. Confluent monolayers of BEAS-2B cells were used. Anchor liposomes (prepared in the same way but without antibody) or a bound irrelevant antibody shows no uptake of liposomal fluorescence (not shown). [for details of immunoliposome preparation and cell culture see [3].

6.9 Further Readings

Handbook of Biological Confocal Microscopy, 2nd Ed., edited by James B. Pawley, Plenum Press, 1995.
This most comprehensive book, covering all fields of biological confocal microscopy and gives a very detailed introduction in the basics underlying this technique.

Cell Biological Applications of Confocal Microscopy, Vol.38 of „Methods in Cell Biology", edited by Brian Matsumoto, Academic Press, 1993.
This book gives an overview of the fundamentals, and focuses on the applications of confocal microscopy.

Molecular Probes Catalogue, actual Ed., Richard P. Haugland, Molecular Probes, Eugene (OR), USA.
This is not only a catalogue, it is the „Handbook of fluorescent probes" with very valuable informations. Deals of course only with Molecular Probes dyes and it should be stated that there are also other providers of fluorescent dyes.

6.10 References

[1] Brakenhoff et al., J. Microsc. (Oxford), 153 (1989), 151-159.

[2] Madry, H., Reszka, R., Bohlender, J., Möller Th., Ganten, D., Wagner J., Renal cell gene transfer by liposomal and other nonviral transfection methods, submitted to Nature Medicine.

[3] Bloemen, P.G. M. et al., Adhesion molecules: a new target for immunoliposome-mediated drug delivery, FEBS letters 357 (1995), 140-144.

Address of the author:
Th. Möller
Max-Delbrück-Centrum für molekulare Medizin
Robert-Rössle-Str. 10
13122 Berlin

7 Introduction to Atomic Force Microscopy and its Application to the Study of Lipid Nanoparticles

Dr. E. zur Mühlen and Prof. Dr. H. Niehus, Humboldt-Universität, Berlin

7.1 Introduction - Traditional Microscopy Techniques

The invention of the optical microscope opened a new world to science. Many advances in material sciences, medicine and biology during the last centuries were only possible with the insight into microstructures, i.e. structures with sizes in the range of 1µm (10^{-6} m). The optical microscope utilises light waves for imaging. In its standard assembly a light beam, defined by a condenser lens system, shines through or onto a sample. The sample is positioned just outside the focal point of the so called objective lens, which generates an enlarged, real image of the sample structure. This real image can be further enlarged via an ocular lens to form a virtual image observable by eyesight. Alternatively the real image can be projected into a camera. Being able to image in any optical transparent medium the light microscope is able to operate in liquids, air and vacuum.

The resolution of an optical microscope is limited by the wave properties of light. Transversal waves, such as light waves, are subject to diffraction. If a wave with wavelength λ propagates through a slit of width d < λ, only part of the light passes straight through. At certain angles relative to this central ray, further rays, which are weaker in intensity, can be found. The intensity pattern behind the slit is called diffraction pattern. According to Abbe's theory of optical microscopy (see for example [1]) next to the central spot at least one further ray of this diffraction pattern has to enter the objective lens in order to form an image. From this theory one can derive the resolution g to be

$$g = 1{,}22 \frac{\lambda}{n \sin \alpha}$$

with n being the refractive index of the medium between sample and objective lens, and α being the opening angle of the objective lens. Thus the resolution can be enhanced by using light of shorter wavelengths (blue light, X-rays) as well as by increasing the refractive index n between sample and objective lens (oil immersion). This formula yields typical resolutions of optical microscopes being of the order of the wavelength of light. Even most sophisticated designs of optical microscopes have resolutions well above 100 nm.

The resolution limits of optical microscopy can be surpassed by probing beams with shorter wavelengths λ. One approach is the use of electron beams. Due to the dualism of wave and matter an electron can be assigned to a wavelength of

$$\lambda = \frac{h}{mv}.$$

Here h refers to the Planck constant, m to the electron mass and v to the speed of the electron. The speed itself can be calculated from the acceleration voltage U of the electron beam to be

$$v = \sqrt{\frac{2eU}{m}}$$

with e being the charge of an electron. For example an acceleration voltage of U = 100 kV yields a wavelength of 0.0039 nm, i.e. a wavelength which is five orders of magnitude lower compared to the wavelength of blue light (~400 nm). This property is utilised in Transmission Electron Microscopy (TEM).

The design of a Transmission Electron Microscope is comparable to an optical transmission microscope. A technical difference is the use of electromagnetic lenses instead of glass lenses. An electron beam is focused by a condenser lens system onto a sample. An enlarged image of the sample is generated by an objective lens. This image is projected by a further lens system onto a fluorescent screen. Calculating the resolution with typical opening angles for this type of microscopes yields a theoretical value of just 0.2 nm, provided an acceleration voltage of U = 100 kV is chosen. Unfortunately spherical and chromatic aberration cause larger problems in electron microscopy compared with optical microscopy, increasing the resolution value by at least an order of magnitude.

The image contrast in TEM is caused by different scattering cross sections for electrons in the sample: some of the beam electrons are scattered at sample electrons, which results in a change of their direction relative to those electrons propagating straight through the sample. The relative number of scattered electrons increases with thickness and electron density of the sample. Thus to guarantee a sufficient contrast while maintaining a reasonable intensity level sample thickness must not exceed a few hundred nanometers (1 nm = 10^{-9} m).

In spite of the large resolution advantages there are some problems which accompany electron microscopy. The main problem is caused by the fact that electron microscopes operate under vacuum conditions to guarantee the unaffected motion of electrons. Quite often, especially for biological samples, this demands a very sophisticated sample preparation. For example all liquid components of an object have to be replaced by vacuum compatible agents. A further problem arises from the fact that the samples have to be conductive to avoid charging of the sample during imaging. Thus samples have to be covered by thin metal films, which may bury small surface features. The electron beam itself may destroy sample areas due to local heating effects. Last not least for good TEM imaging conditions samples have to be very thin requiring sophisticated cutting instruments. A recent overview over biological sample preparation techniques can be found in [2].

These difficulties in sample preparation are as well valid for *Scanning Electron Microscopy (SEM)*, which uses an entirely different mechanism for imaging. In a scanning electron microscope an electron beam scans a sample. The electron beam is well collimated, i.e. it has typical diameters of a few nm, and is accelerated by a voltage in the order of magnitude of 10 kV. As the electron beam hits a sample its electrons transfer their energy to sample electrons. Some of these electrons, called secondary electrons, are able to escape from the surface of the sample. The amount of emitted electrons is dependant on the electron density of the sample close to its surface as well as to the surface area exposed to the sample. The low energetic secondary electrons are collected by a detector above the sample. An image is generated by mapping the number of secondary electrons vs. the position in a raster. Its resolution is determined by the size and energy of the electron beam as well as by the chemical nature of the sample, and typically does not decrease below 10 nm.

A further surface sensitive imaging technique - Scanning Probe Microscopy - has the abilities to overcome these sample preparation problems and to increase imaging resolution on many samples. Following in this chapter this technique and some of its subdivisions will be introduced.

7.2 Introduction - Scanning Probe Microscopy

In an abstract definition *Scanning Probe Microscopy (SPM)* implies, that a sharp tip is brought either in contact or in the immediate vicinity of a surface, probing surface properties with a high spatial resolution. An image is acquired by stirring the tip parallel to the sample surface, or alternatively, the sample surface parallel to the tip, while acquiring these properties at selected coordinates. Typically a square or rectangular array of coordinates, called a *raster*, is chosen.

Due to the variety of surface properties to be probed SPM divides into various subsets. *Scanning Tunneling Microscopy (STM)* probes the density of electronic states at a surface. It is based on the acquisition of currents between a metal tip and a conductive surface. *Atomic Force Microscopy (AFM)*, which is also called *Scanning Force Microscopy (SFM)*, maps surface forces acting between a sample surface and a tip, which is attached to a cantilever. Depending on the particular forces to be acquired

AFM may divide into further 'microscopies', some of those will be discussed later in this article. As no electrical currents are involved during data acquisition the sample surface may be insulating, suggesting this technique often as the better choice for biological or pharmaceutical samples. Thus AFM received growing attention in this field during the past few years, accompanied by the development of suitable sample preparation techniques necessary for molecular scale imaging (see [19] and [20], for example).

Besides STM and AFM there are many more SPM techniques. For example *Scanning Near-Field Optical Microscopy (SNOM)* combines the measurements of surface forces with the acquisition of optical properties and luminescence by using an illuminated glass-fibre as probing tip. It is an optical microscopy technique capable of surpassing resolution limits imposed by the wave properties of light. Due to the large variety of the probed properties a discussion of SNOM and other SPM techniques goes beyond the scope of this article and the reader is referred to [3]. Following the concept of STM will be explained, before emphasis will be put on AFM and its various techniques.

7.3 Imaging Mechanism of STM

STM is the oldest of all SPM techniques, and more than one decade ago the first tunneling microscope has been realised by Binning and Rohrer [4]. The operation of STM is based on a quantum mechanical effect called electron tunneling.

Consider an electron caught within the boundaries of a potential barrier. For example, free electrons in a sample or a tip experience their surfaces as boundaries. According to the classical concepts of physics an electron can only overcome such a potential barrier if the potential threshold is lower than the electron energy. However the physics of atomic dimensions are subject to the laws of quantum mechanics. It can be derived from these laws that there is a certain probability for an electron to surpass the walls of a quantum well, even if its energy barrier is higher than the energy of the electron. This probability decreases exponentially, i.e. very rapidly, with distance from the quantum well. Within a few atomic radii already it may be assumed to decrease to zero.

Thus if a metal tip is brought into the immediate vicinity (~0.5 nm) of a conducting sample surface the areas, where electrons can be found outside tip and sample, overlap. There is a permanent exchange of electrons between tip and sample, although tip and sample are not in direct (ohmic) contact. At the same time the Fermi levels, i.e. the maximum energy levels to which electronic states are filled, of tip and sample are pinned to the same energy value. As a result the net current between tip and sample is zero, i.e. the number of electrons passing from tip to sample is equal to the number penetrating into the opposite direction during the same time unit.

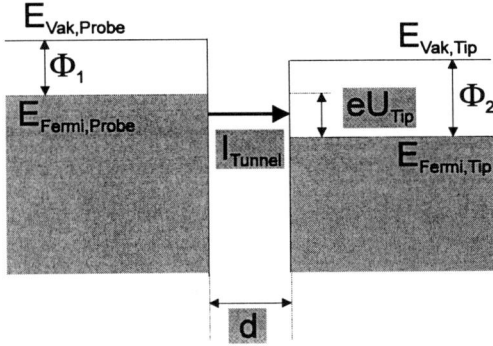

Fig. 7-1: Displayed is the principle of electron tunneling between a metal tip and a metal surface. The left part of the image shows an energy diagram of the sample, the right part the according diagram of the tip. The electronic states are filled up the respective Fermi levels, which yield an almost equal energy level upon approach of tip and sample. If a negative voltage is applied to the sample, the Fermi level of the sample is shifted with an energy of eU_{Tip} relative to the Fermi level of the tip. A quantum mechanical law allows a flux of electrons from the filled sample electronic states into the vacant tip electronic states even if tip and sample are not in immediate contact. The tunneling current is strongly dependant on the distance between tip and sample.

As soon as a voltage is applied between tip and sample the Fermi level of the tip is shifted with respect of the Fermi level of the sample. For example if a small, negative voltage is applied to the sample vs. a grounded tip, the sample's Fermi level is raised vs. the Fermi level of the tip (see Fig. 7-1). Consequently some occupied energy levels of the sample are equal in energy to some vacant levels in the tip causing a net tunneling current of electrons flowing from sample to tip. Alternatively if a small, positive voltage is applied to the sample there is a tunneling current from tip to sample. The discussion becomes more complicated if semiconductor samples are investigated. Now the Fermi level of the tip has to be raised to the lowest level of the conduction band to allow any electrons to flow from tip to sample. Alternatively it has to be lowered below the upper edge of the valence band until the first valence electrons are able to tunnel to the tip.

A more detailed theoretical discussion [5] for metal samples yields for the functional relation between tunneling current I_{tunnel}, voltage between tip and sample U and tip to sample distance d

$$I_{tunnel} \propto \exp(-cd).$$

c is a positive constant and depends for a given tip and sample system on the average of electron energies needed to escape from tip and sample, i.e. the average of the work functions Φ_{tip} and Φ_{sample}. The relation yields a very strong dependency of the tunneling current on tip to sample distance for a given voltage. In fact considering typical values for the work functions a change of 0.1 nm in distance between tip and sample causes a change of an order of magnitude in tunneling current. Thus the acquisition of tunneling currents allows tip to sample distance measurements in atomic dimensions.

In a standard design the metal tip, which is cut from a metal wire, is mounted onto a piezo tube allowing accurate positioning in the sub-nm range. While the tip is moved across the sample changes in topography cause changes in tunneling current which are acquired to serve as input to an electronic feedback loop. The feedback loop causes the piezo to correct the tip position until an originally set current value is obtained. By this mechanism the tip is able to follow topography features on the surface, and an image of the surface can be obtained if the changes in vertical tip position are monitored vs. the position parallel to the sample surface. Technical details will be described later in larger detail when AFM operation will be discussed.

In STM of metal samples typical applied voltages between tip and sample are of the order of several 10mV yielding tunneling currents in the nA ($1nA = 10^{-9}A$) range of tip to sample distances of 0.5 to 1 nm. Imaging semiconductor samples voltages have to be high enough to overcome the band gap and are in the 1V range to yield similar distance and tunneling current values. Due to the lack of free electrons in the sample it is generally impossible to operate STM on insulating samples.

It should be noted that STM does not image topography information directly. As described STM operation is based on the tunneling of electrons between an occupied and an unoccupied electronic state and thus, in physical terms, STM images the density of electronic states rather than pure topography information. For example atoms with a high electron affinity, such as oxygen, cause a decrease in tunneling current relative to neighbouring metal atoms. During scanning this decrease in current causes that the feedback loop lowers the tip. Consequently in atomic resolution imaging on metal surfaces oxygen atoms are usually monitored the same way as depressions in surface topography [6].

Being the first scanning probe microscopes widely available tunneling microscopes were used in imaging biological samples at an early stage. For example DNA was imaged successfully more than seven years ago [7, 8]. Despite many successes in high resolution imaging of molecules the application of STM to biological samples in general is limited owing to its demand for conducting samples. By far the most pharmaceutical or biological samples show a low conductivity. To become suitable for STM imaging these samples have to be covered by a thin metal film introducing many complications to the sample preparation process. Thus ever since atomic force microscopes became more widely available about four to five years ago the main focus for imaging biological samples shifted to this technique.

7.4 Forces Acting between Probe and Sample

The operation principle of an atomic force microscope is based on the measurement of interatomic forces between a sharp probing tip and a surface. A typical tip imaged by SEM is shown in Fig. 7-2. The silicon tip has a height of about 12 µm and is shown standing upright on a cantilever, which measures 220 µm in length and about 25 µm in width. The inset displays an ontop view onto the tip end with its threefold symmetry. The tip remains sharp within the display limit, higher resolution magnification images reveal that the tip end has rather spherical shape with diameters typically ranging between 10 to 100 nm. Higher resolution TEM images of metal tips show that these 'half-spheres' consist of small atomic scale tips emerging from the tip surface [19].

The various strengths of the forces involved depend on the nature of the sample, the tip material and geometry as well as on the distance between probe and sample. In air, contamination of the surface causes additional forces which have to be considered as well. A thorough theoretical discussion of these forces and their possible interactions on an atomic scale is rather difficult owing to the fact that a very large number of tip

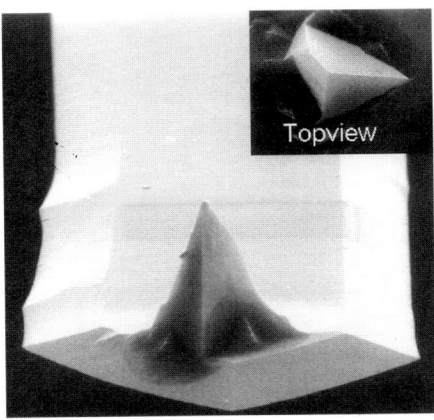

Fig. 7-2: Shown is a sideview SEM image of a commercially available tip used in non-contact AFM operation. The tip has a height of 12 µm and can be seen standing upright on a cantilever. The cantilever has a length of 220 µm and a width of 25 µm. The inset shows an ontop view onto the tip revealing its threefold symmetry.

and sample atoms are involved. Nevertheless the natures of the individual forces are well understood and, in the following, these forces will be discussed separately.

For the case of simplicity consider two neutral, non-polar atoms in the gas phase (see Fig. 7-3). Separated by a large distance of at least several ten nm, these atoms do not exchange any forces. On approach the atoms experience an attractive force first. Atoms are always subject to fluctuations of their charge density, i.e. there is an instantaneous displacement of the negatively charged electron cloud relative to the positively charged nucleus of the atom. Thus the atom becomes a dipole for a moment, which may induce a shift of the electron cloud relative to its nucleus in a second atom. This results in an attractive force between the two atoms, known as *van der Waals* force. The strength of the force is proportional to about $1/d^8$ for distances d larger than 10 nm between the atoms and $1/d^7$ for shorter distances. Thus the van der Waals force decreases rapidly with increasing distance d and is only effective over distances below several 10 nm.

As the distance between the two atoms decreases further their electron clouds begin to interact directly. Firstly an overlap of the electron shells results in their respective displacement outwards relative to the atom nuclei. This causes an incomplete shielding of the charge of the atom nuclei and thus leads to the exchange of repulsive Coulomb forces. Secondly due to the Pauli exclusion principle, two electrons with equal quantum numbers are not allowed to coexist in the same system. Thus the respective electron states can only overlap if the quantum mechanical state of one of the electrons

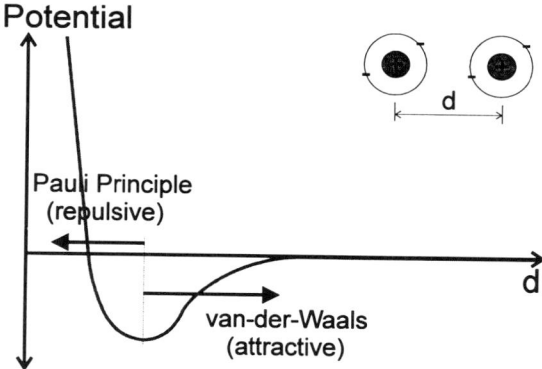

Fig. 7-3: Displayed is the interaction potential between two individual, single atoms as a function of distance between the atoms. Upon approach the atoms experience the weaker, attractive forces (such as van der Waals forces) before the overlap of electron clouds causes strong repulsive forces, which result in the steep slope in the left part of the curve.

is changed, which typically requires the input of energy yielding in an additional repulsive force. In summary as the interatomic distance d decreases to values which are in the range of atomic radii (few tenth of a nm) the atoms exchange strong *interatomic repulsive forces*, which entirely dominate the van der Waals forces. A quantitative analysis of exchanged forces leads to values of several nN (1nN = 10^{-9}N) and above.

Mathematically the shown dependence between force and interatomic distance summarised in Fig. 7-3 can be derived from the Lenard-Jones potential:

$$V(d) = -3E_{Eq}\left[\left(\frac{\sigma}{d}\right)^{12} - \left(\frac{\sigma}{d}\right)^{6}\right]$$

with E_{Eq} being the lowest potential energy at the equilibrium distance $d_{Eq} = 2^{1/6}\sigma$. The force shown in Fig. 7-3 can be gained by calculating the derivative of V(d) vs. d. It should be kept in mind, that the Lenard-Jones potential is a rough approach to any given tip and sample system. For a complete description the interaction between many tip and sample atoms have to be considered. A complication arises especially due to the fact that van der Waals forces acting in an array of atoms are not simply the sum of van der Waals forces acting between two single, isolated atoms from this array. Instead the presence of a momentary dipole in one atom induces dipoles in all surrounding atoms causing a change to their interaction forces as well. A detailed review of theories on van der Waals forces in terms of solid state physics as well as some new approaches can be found in [9]. In this study a calculation of van der Waals forces acting between a sphere with radius R (in approximation of a spherical tip) and a flat sample surface yields for a sphere to sample distance d of a few nm

$$F_{vdW}(d) = -\frac{H}{6}\frac{R}{d^2}$$

with H being the so called Hamaker constant, a function of the refractive indices of sample, sphere and immersion medium as well as on the absorption energies of these media. It should be noted that as a result of this more detailed model the force decreases with $1/d^2$ and thus slower compared to the distance dependence predicted by the Lenard-Jones potential. A calculation of typical van der Waals forces yields forces in the pN (1pN = 10^{-12} N) regime, being about three orders of magnitude lower than the repulsive interatomic forces.

Depending on the nature of sample and tip there may be additional forces exchanged between tip and sample. For example, if a grounded metallic tip approaches an electrically charged sample area, an image charge of equal amount but opposite sign will be induced at the tip. Thus there will be an additional attractive *electrostatic force* acting between tip and sample. The strength of this force can be calculated from

$$F_{el}(d) = q_{Tip}E(d)$$

where q_{Tip} is the electrical charge induced on the tip and E(d) is the electrical field generated by the sample charge at a distance d above the sample. In a similar way magnetic domains on a sample surface will influence a ferromagnetic probing tip. Depending on tip polarisation this *magnetic force* can be either attractive or repulsive. Magnetic and electrostatic force are both far ranging forces and are even exceeding the range of van der Waals forces.

The range of forces determines the effective tip size, i.e. the part of the tip, which interacts with the surface. The effective tip size is of fundamental importance to AFM, as it determines the imaging resolution.

The exchange of repulsive interatomic forces is limited to the interaction between a sample surface atom and the most protruding tip atom. As the interaction is thus limited to two single atoms atomic resolution in force mapping can be achieved. In contrast van der Waals forces are longer ranging and thus larger tip areas interact with the surface. As it decays with $1/d^2$ with distance from the surface though, the most protruding end part of the tip dominates force exchange. Furthermore van der Waals forces can only be probed at distances above a surface where they are not superimposed by repulsive forces, i.e. typically at distances larger than 1nm. Consequently force exchange between tip and surface involves larger areas, and in force mapping the lateral resolution typically decreases to about 1nm. The same arguments apply to the exchange of electrostatic and magnetic forces. Being further ranging the best possibly obtainable resolution in probing these forces with a tip is even lower.

Studying forces between tip and sample, which are placed in an ambient air environment, a further complication because of the appearance of *capillary forces* has to be considered. In ambient air any non hydrophobic surface is covered by a thin contamination layer, which mainly consists of condensed water molecules. The absolute thickness of such a layer reaches values of up to 20 nm, depending on surface properties such as curvature and hydrophobic character as well as on air humidity. Thus on approach to the sample surface the contamination layers of tip and surface overlap before tip and sample are in immediate contact. As soon as the layers overlap the system will try to minimise its energy by decreasing its total surface area. This causes a strong capillary force pulling the tip towards the surface. Attractive forces develop as well in a humid environment, even if tip and sample are hydrophobic. This is due to spontaneous capillary condensation of vapour in the cavity opening between tip and surface next to the contact area, forming a liquid bridge. Regardless of the scenario the strength of the capillary force can be calculated from thermodynamic equilibrium considerations involving the Kelvin equation, which relates the shape of the meniscus to the relative vapour pressure [10]. Then the attraction between tip and sample yields

$$F_{Capillary}(d) = \frac{\pi RT\rho}{M} \ln\left(\frac{p}{p_s}\right) r(t-d)$$

with 2r being the diameter of the water bridge between tip and sample which is about equal to the tip radius, R the universal gas constant, T the temperature, ρ and M the mass density and the molar mass of the wetting liquid, p/p_s the relative vapour pressure, which, in ambient air, is equal to the relative humidity, t the thickness of the contamination layer and d the distance between tip and sample (from [9]). The capillary force is about proportional to the tip radius, and consequently a sharp tip will experience lower capillary forces compared with a blunt tip. A calculation for a typical tip radius of 50 nm yields a capillary force of about 50 nN. Thus capillary forces are a few orders of magnitude larger than van der Waals forces. Nevertheless it is still possible to separate capillary forces from van der Waals forces by their different quality of force increase with decreasing distance between tip and sample.

An unwanted side effect can be found in the fact that capillary forces limit the minimum repulsive forces, because they pull a tip into contact with a sample until they are equilibrated by repulsive force. Promising approaches to solve this problem can be found by either operating the system in a dry gas atmosphere, or by complete immersion of tip and sample into a liquid.

7.5 Contact-AFM: Topography Imaging

In a force microscope the probing tip is attached to a cantilever type spring. Standard cantilevers which are commercially available for contact imaging, i.e. imaging in the range of repulsive interatomic forces, have spring constants of k = 0.032 nN/nm. For example a 0.1 nN change in force, a value below typical values for repulsive interatomic forces, results in a bending of a free tip cantilever system of about 3 nm, a distance large compared to atomic corrugations. As a result if such a tip is brought in contact with a surface the tip is generally able to be stirred across a surface following surface features without destroying or otherwise altering a sample surface.

Prior to imaging the tip is brought into contact with the surface, i.e. tip and sample exchange interatomic repulsive forces. As the sample is then moved laterally (parallel to the sample surface), the tip follows topographic features resulting in a bending of the cantilever. Thus to produce an image of the surface changes in cantilever position have to be monitored. This can be a very delicate task regarding the fact that changes in cantilever deflection may be as small as atomic corrugations (<0.1 nm). There are

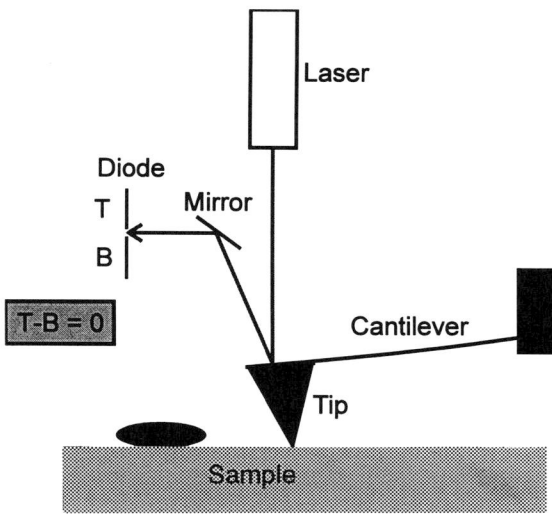

Fig. 7-4: Basic components of a force microscope (not to scale). A tip is held by a soft cantilever and touches the sample surface. The vertical tip position is acquired by an optical deflection method: a laser beam is reflected from the cantilever, and the position of the reflected beam spot in a photodiode is monitored as photocurrent difference between the top and bottom part of the photodiode. Prior to data acquisition the instrument is aligned to yield a previously chosen value for the photocurrent difference (e.g. T-B = 0).

several approaches in design of atomic force microscopes, namely the tunneling method [11], laser interferometry [12, 13] and optical position sensitive method [14]. Owing to its reproducibility, versatility and relative ease of operation the optical method is the by far most commonly used and thus will be described further.

The optical position sensitive method magnifies cantilever deflection changes by monitoring the position of a reflected laser beam from the cantilever (Fig. 7-4). In a typical design a diode laser is placed above the cantilever. The focused laser light is aligned onto the free end of the cantilever, which is mounted with a tilt of a few degrees downward relative to the horizontal plane. Thus the laser beam is reflected from the cantilever under twice this angle relative to the incoming beam. The reflected beam propagates via a mirror into a position sensitive photodiode. This diode consists of four quadrants, two placed at its top (abbreviated T in Fig. 7-4) and two placed at its bottom (abbreviated B). Initially mirror and photodiode are aligned to place the reflected laser beam in the centre of the four quadrants, so that the difference in photocurrent between top and bottom diode is equal to a prechosen value, called setpoint (chosen to be zero for simplicity in figure Fig. 7-4). Thus as the sample is

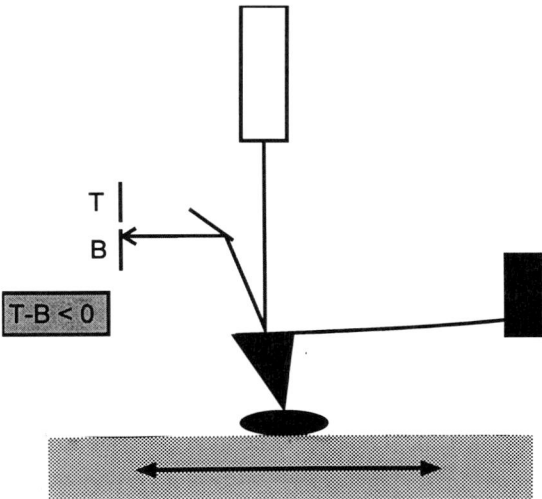

Fig. 7-5: Response of the force microscope to topographic changes during scanning: in the example the tip is lifted resulting in a shift of the reflected beam towards the bottom part of the photodiode. The change is monitored by a feedback loop, which causes a retraction of the surface until the previous value of photocurrent difference is monitored again.

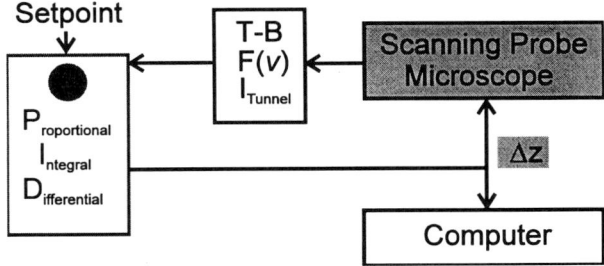

Fig. 7-6: Feedback-loop mechanism controlling the vertical sample position. The actual photocurrent difference signal T-B is compared to a previously chosen one, called 'setpoint'. The difference between the values is used by a Proportional-, Integral- and Differential controller (PID-controller) to calculate a new vertical sample position (Δz), which is fed back to the microscope and simultaneously stored as heightvalue in a computer. Exactly the same device is used in STM or non-contact AFM to keep tunneling currents (I_{tunnel}) or oscillation amplitudes ($F(v)$) at a constant value.

moved laterally relative to the tip changes in topography cause a shift of the laser spot at the photodiode, which is electronically accessible as change in photocurrent difference between top and bottom quadrants of the diode (see Fig. 7-5).

There are two distinct possibilities to process the photocurrent difference. In the so called *constant force mode* of AFM operation this signal serves as input into a feedback loop (see Fig. 7-6). In a first processing step the difference Δ between the actual photocurrent difference value and the setpoint is calculated. This difference is used by the loop to calculate a new vertical sample position. The sample is raised or lowered to this new position, respectively, and again the photocurrent difference value is taken. The loop is passed repeatedly until the level of the setpoint is reached again. Height information is directly accessed by monitoring the total change of vertical sample position before the sample is moved laterally to a new acquisition point. Thus in an ideal situation, in which the feedback loop operates without any time delay, cantilever deflection and consequently the force acting between tip and sample is kept at a constant value. In a real force microscope the loop needs some time to respond. However if the speed of lateral sample motion is chosen to be low enough the influence of the finite response time of the loop can be negligible and the system responds almost ideally. Thus to operate a force microscope in the constant force mode successfully scan speed and parameters controlling the response of the feedback loop have to be adjusted carefully for each sample.

In *constant height mode* AFM operation the changes in photocurrent difference are monitored directly while the vertical position of the sample remains unchanged throughout the acquisition process. There is no feedback mechanism involved and consequently there are no complications arising from its finite response time allowing faster data acquisition. Some serious disadvantages accompany constant height mode operation, though. In this mode the system is not able to follow large changes in topography resulting in an excessive bending of the cantilever and large change in forces acting between tip and sample. Furthermore real height information is not easily accessible. During imaging only photocurrent differences are monitored. The calculation of real height values from these currents is rather undefined, as they depend as well on system alignment and optical cantilever reflection. These parameters are difficult to obtain and may change during the course of work. Concluding it may be stated that the constant height mode can only be chosen for AFM operation if an almost atomically flat sample is to be imaged and pure surface geometry instead of total height information is to be accessed. For all other samples and imaging conditions the constant force mode should be preferred, and is therefore most commonly used even inspite of more difficult handling procedures.

A topography image is obtained by acquiring the total vertical change in sample position (or alternatively the changes in photocurrent difference) vs. the lateral position of the sample relative to the tip. An image displayed in Fig. 7-7 is taken by moving the sample in a raster. The height information is mapped in a square array of coordinates in the sample plane and stored in a computer. For display purposes height values are translated into a linear greyscale. Dark greyvalues represent low heightvalues while bright greyvalues are assigned to higher areas. The resolution of the image is then simply determined by the distance between neighbouring coordinates, as long as other resolution boundaries discussed in the section about tip to sample interactions do not apply. Typically the number of datapoints ranges between 200x200 and 1000x1000 per image.

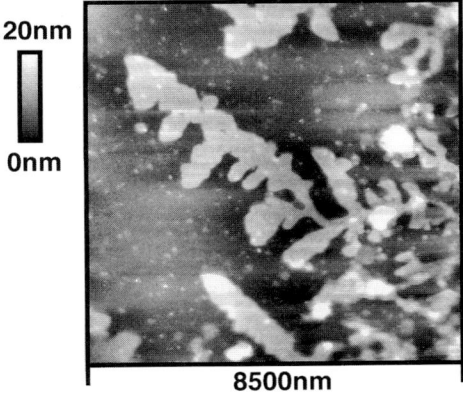

Fig. 7-7: Topography image of dendritic plates growing from SLN on hydrophobic silicon surfaces within days of storage. The image covers a sample area of 8500 nm x 8500 nm. The greyscale covers height differences of 18 nm, with high areas appearing in lighter, low areas in darker greyvalues.

Shown in Fig. 7-7 are dendritic plate structures growing from SLN, which consist of a fat core surrounded by a surfactant layer [15], on freshly etched silicon surfaces. The scanarea covers a surface area of 8500 x 8500 nm^2. The plates appear after a few days of storage in an ambient air environment, and it is proposed that the plates form from surfactant molecules migrating off the lipid nanoparticles. With AFM it is easily possible to measure the height of the plates relative to the silicon substrate level to be 7 nm.

So far the discussion of resolution has been limited to tip to sample interactions and storage capacity (i.e. the number of datapoints). It has been assumed so far throughout this article that the positioning accuracy of the sample is adequate not to limit the best possible resolution in any way. As a matter of fact this problem was - next to the design of suitable tip cantilever systems - one of the most difficult parts to overcome. A suitable approach had been found in the use of piezoceramic elements. These ceramics alter their shapes if placed into an electric field. For example consider a thin tube ceramic, with a grounded electrode at the inside and a second electrode at the outside of the tube. If a potential is applied at the outward electrode a given piezoceramic tube will expand or contract linearly - depending of the sign of the potential. Motion perpendicular to the tube axis can be gained by placing two separate electrodes at the outside of the tube. If one potential is chosen to be positive, the other one selected to be negative the tube will expand at one side and contract at the opposite side resulting in bending of the tube. It is possible to bend a tube in all direction if four equal electrodes are placed around the tube. In addition the tube can change its length by applying a further bias voltage to all electrodes at the same time. Thus if a sample is placed ontop one end of such a piezoceramic tube it can be positioned laterally and perpendicularly relative to its surface. The positioning accuracy for these so called tube scanners reaches values around 0.01 nm and thus does not limit the resolution of a force microscope.

If large sample areas are to be scanned the use of tube scanners is disadvantageous. Non-linear components in piezoceramic response to the applied voltage become a problem for large ceramic distortions. Furthermore due to the bending during lateral motion the sample surface is not strictly stirred in a plane. In fact it rather follows a sphere segment. Negligible for small imaging areas this causes serious problems if large areas on very flat samples are to be visualised. For these reasons so called tripod scanners are chosen for imaging areas above $10 \mu m \times 10 \mu m$. In this design three piezoceramic are mounted perpendicular to another. Each ceramic is able to move in one direction only; two are responsible for lateral motion while the third piezo raises or lowers the sample.

Technically further complications arise due to vibrational noise and thermal expansion during imaging. Any noise source - possibly ranging from low frequency oscillations within a building construction up to high frequency acoustic or electronic noise - causes unwanted motion between tip and sample surface possibly which may result in large displacements on the imaging scale. A further disturbing source can be found within the expansion of parts within the microscope due to temperature changes. Again an expansion of the microscope assembly may result in disturbing motion. Last not least the cantilever may be excited by thermal noise [16] causing small oscillations parallel to the sample surface.

Finally in Fig. 7-8 all the components discussed so far are summarised. Laser, tip, mirror and photodiode should not change their relative positions during imaging and thus are mounted into a solid block shown at the top. The sample is placed ontop the piezoceramic tube. The tube itself is fixed on a translation stage allowing precise sample positioning prior to imaging. Sample positioning as well as coarse approach between tip and sample are controlled by a video camera (not shown in the figure). To avoid vibrational influence on the system during imaging the whole assembly is held by long viton-rings or springs and should be placed on a damping table for high resolution imaging. In comparison with electron microscopes atomic force or scanning tunneling microscopes are rather compact in design. The displayed assembly is not larger than a fist.

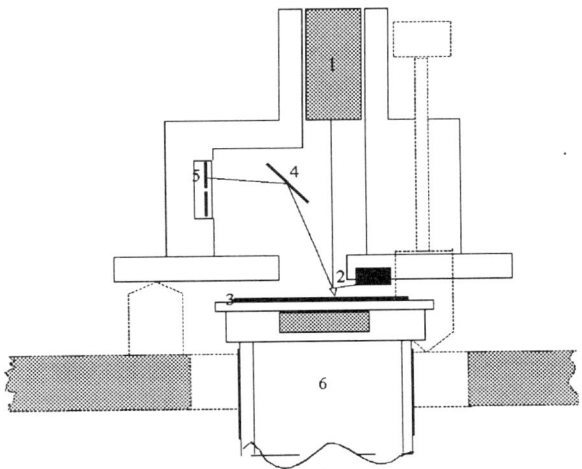

Fig. 7-8: Principle set-up of a commercial force microscope [21]. Laser diode (1), cantilever and tip (2 - not to scale), mirror (4) and four-zone photodiode (5) are mounted in a head assembly. The probing tip interacts with the sample (3), which is placed ontop a piezoceramic tube (6), which alters the vertical sample position while scanning the sample in a raster.

7.6 Contact-AFM: Lateral Force Microscopy (LFM)

In lateral force microscopy (LFM) the tilt of the tip in scan direction is acquired giving access to frictional information on a nanometer scale. As the sample is moved relative to the tip frictional forces are acting between tip and sample. The frictional force vector points antiparallel to the vector of tip to sample velocity, acting on the end of the tip which is in contact with the surface. This causes a lag of the tip end relative to the cantilever, and a distortion of the cantilever itself. The physical nature of the frictional forces can be quite complex. Next to surface roughness the reactivity and chemical nature of the surface influences the frictional force as well [17]. Thus changes in lateral force reveal different surface qualities and are helpful in distinguishing between different species on, or within a sample surface.

During lateral force imaging the force microscope operates in contact mode. Next to topography mapping the tilt of the tip is acquired by monitoring the distortion of the cantilever parallel to its long axis (see Fig. 7-9). The distortion causes a 'left-right' shift of the laser spot in the photodiode in addition to the 'top-bottom' motion of the spot used for contact topography imaging. Thus by acquisition of the photocurrent difference between the left and right quadrants of the photodiode a lateral force image is obtained at the same time as the topography is mapped in contact mode.

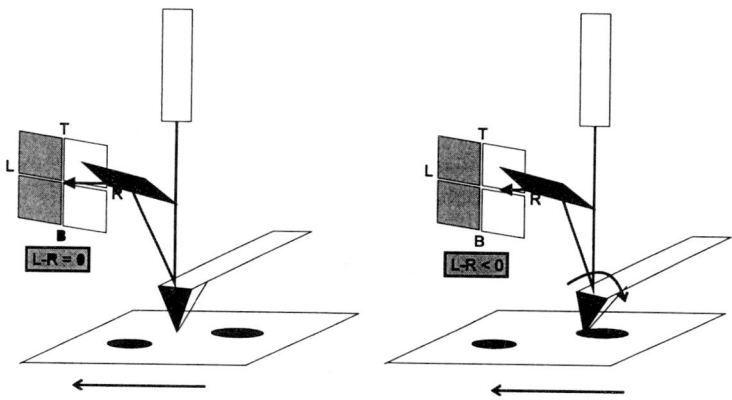

Fig. 7-9: Principle of LFM. Low frictional forces result in a small cantilever tilt during scanning (left). Increased friction results in larger distortions parallel to the cantilever axis (right). The distortion can be acquired by monitoring the photocurrent difference between the left and right quadrants of the photodiode.

Fig. 7-10 displays a lateral force image of the same structure which is shown in the topography image of Fig. 7-7. The greyscale represents the 'left-right' photocurrent difference directly. Dark areas are assigned to low differences related to small values of tip tilt and frictional force. Brighter areas indicate larger tilts and therefore larger frictional forces. In Fig. 7-10 the dendritic plate structure appears dark pointing at low lateral forces in this area. In contrast the lateral force signal on the surrounding parts is higher and quite noisy indicating a high reactivity of the surface and larger frictional forces. This should be expected from a rough, hydrophobic silicon surface. During the

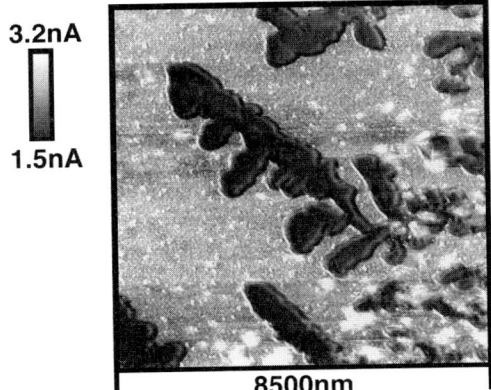

Fig. 7-10: LFM image of the same sample area displayed in Fig. 7-7. Dark greyvalues represent small distortions, while bright greyvalues are assigned to large tilts of the cantilever parallel to its long axis. The large contrast between the dendritic plates and the surrounding silicon surface reveal the different chemical nature of the plates.

discussion of Fig. 7-7 it was proposed that the plates consist of surfactant molecules while the displayed surrounding area was assigned to the silicon surface. The lateral force data supports this assumption.

7.7 Contact-AFM: Force-Distance Curves

The small spring constant of cantilevers used in contact operation AFM has further important consequences on the way the tip will react on tip to sample forces. This behaviour is summarised in Fig. 7-11, where the bending of the tip end of the cantilever vs. the relative position between the free tip-cantilever system and a silicon sample is displayed. Generally the force $F_{Cantilever}$ on the cantilever itself can be calculated from Hooke's law:

$$F_{Cantilever} = k_{Spring} z$$

with k_{Spring} being the spring constant of the cantilever (e.g. 0.032nN/nm) and z being the bending of the cantilever.

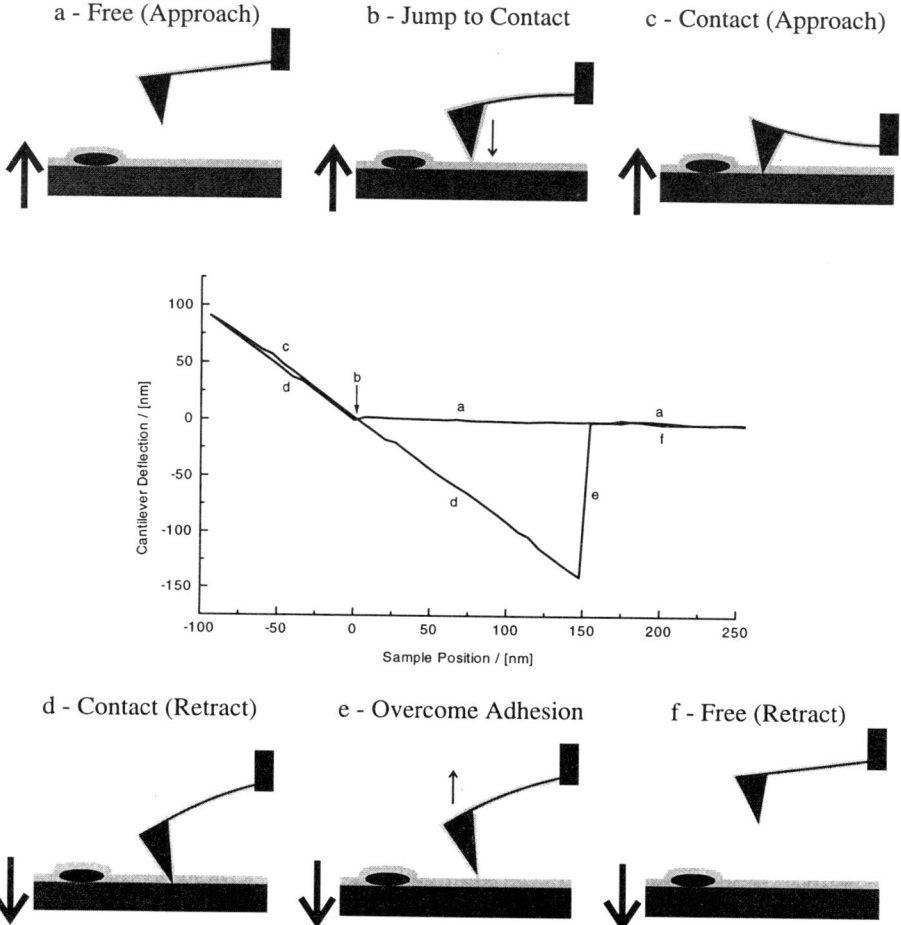

Fig. 7-11: Force-distance curve obtained with a standard tip on a silicon surface. The x-axis displays the vertical sample position, the zero value is chosen to match the 'jump to contact' position. The y-axis displays the bending of the cantilever. From this value the force acting on the cantilever can be easily calculated using Hooke's law.

With a tip far above the surface there is no force exchange and thus no influence on the cantilever (horizontal line labelled (a) in Fig. 7-11). The capillary force pulls the tip rapidly onto the sample surface as soon as the contamination layers of tip and surface overlap (small dip labelled (b) in Fig. 7-11). Alternatively in liquids, under vacuum or under dry protective gases van der Waals forces are responsible for this sudden 'jump to contact' as soon as these forces are large enough to destabilise the tip-cantilever system, i.e. as soon as the force gradient is larger than the spring constant of the cantilever. Once tip and sample are in contact they exchange repulsive interatomic forces, and the tip follows an upward motion of the sample exactly (increasing deflection with decreasing sample position on the left hand side, labelled (c), in Fig. 7-11). The total value of the force acting between tip and sample is the sum of capillary force $F_{Capillary}$ and the force caused by the bending of the cantilever $F_{Cantilever}$. The upward motion of the tip is reversible upon retraction of the sample (upper part of line labelled (d) in Fig. 7-11) if there is no inelastic deformation of the sample involved. However as soon as the former 'jump to contact' position is surpassed the tip still remains in contact with the surface as it is held by the capillary forces (lower part of line labelled (d) in Fig. 7-11). The cantilever is bent downwards until the resulting force $F_{Cantilever}$ overcomes the adhesive forces. As soon as the tip is free the cantilever snaps away (line (e) in Fig. 7-11) and rests in its original position (line (f) in Fig. 7-11) after a short oscillation period.

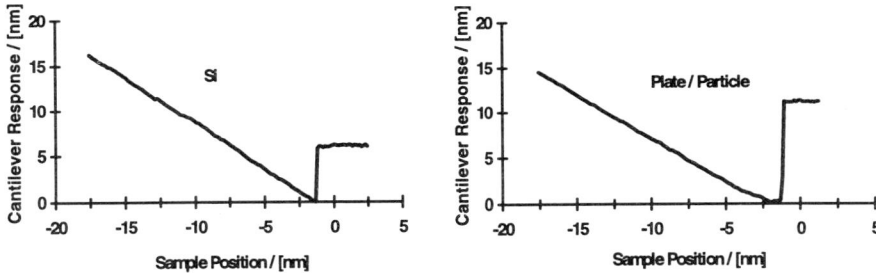

Fig. 7-12: Force-distance curves obtained on silicon (left) and SLN (right). Shown are only a small section around the 'jump to contact' region as compared to figure Fig. 7-11. The tip is pulled into contact 6nm above the silicon surface, compared to 11 nm above a SLN. This indicates that a thicker contamination layer covers the SLN, pointing at their hydrophobic properties. Upon contact the force-distance curve of the SLN shows a delayed response, in contrast to the curve obtained on silicon. This is interpreted as an initial deformation of the outer surfactant layer surrounding the fat core of the particle (from [15]).

Force distance curves reveal elastic properties of a sample. In ambient air operation additional information about the quality of the contamination layer can be gained. As an example the jump to contact regions of two force distance curves are displayed in Fig. 7-12. The top curve shows the cantilever response during approach on a freshly etched, hydrogen terminated Si(111) surface. The surface is known to be hydrogen passivated and thus hydrophobic [18]. Before contact the cantilever does not respond to the approaching surface. Thus the response curve in the diagram proceeds horizontally. At a distance of 6nm between tip and surface, the contamination layers overlap pulling the tip rapidly onto the surface, which results in a vertical course of the response curve. As the sample is not deformed by interatomic repulsive forces, an assumption which is reasonable on crystalline silicon, the tip follows the upward sample motion immediately. Further response is equal to sample motion, resulting in the straight line in the left section of the response curve.

The response curve in the bottom part Fig. 7-12 has been obtained on a drug loaded solid lipid nanoparticle (SLN), which is placed on the silicon sample. The particle with a diameter of about 150 nm consists of a crystalline fat core, which is surrounded by a soft surfactant layer. The particles and their properties as characterised by other techniques are described in [15]. Due to the surfactant layer the particle has a hydrophilic character, and one would expect that a thicker contamination film surrounds such a nanoparticle compared to hydrophobic silicon. Indeed already at a tip to sample distance of 11 nm the contamination layers of tip and surface overlap pulling the tip into contact, as can be measured from the vertical jump to contact section from the response curve. Thus the contamination layer around a hydrophilic particle is 5 nm thicker as compared with the one ontop a hydrophobic silicon surface. It should be stressed that the absolute thicknesses of contamination layers are not directly accessible as the thickness of the layer covering the tip is not known. Furthermore absolute values depend on different parameters such as the daily air humidity and thus are not easily reproduced. Nevertheless with the acquisition of relative changes in layer thickness it is thus possible to distinguish between hydrophobic and hydrophilic characters locally. Once in contact there is a delay of 2 nm between tip and sample motion, i.e. the tip deforms the particle by about 2 nm before it follows the sample motion almost equally. It may be stated that the delay is only due to an elastic deformation of the surfactant layer surrounding the particle as it appears as well on surfactant plates placed on silicon. Furthermore the delay does not alter with changing particle diameters as would be expected if it is caused by a deformation of the core itself. Thus it is conceivable that once the forces stored in the deformed surfactant layer balance the capillary forces the tip follows the sample motion almost equally with small deviations being caused by further deformations due to the increase in interatomic repulsive forces.

Force distance curves obtained from the dendritic plates displayed in figure Fig. 7-7 reveal, that the plates have a hydrophilic character and show the same elastic properties as SLN. Therefore it is conceivable that the plates form as surfactant molecules migrate from the shell of the particle onto the silicon surface.

7.8 Contact-AFM: Scanning Force Spectroscopy (SFS)

In Scanning Force Spectroscopy (SFS) imaging the elastic properties of a sample are revealed. In addition to normal topography imaging in contact mode the sample is oscillating at a relatively low frequency (a few kHz or cycles per datapoint) and with an amplitude of a nm and below. Consequently the force between probe and sample is varied around the force value as chosen by the setpoint. The respective response of the cantilever is monitored as amplitude in photocurrent difference between top and bottom part of the diode and is acquired in units of photocurrent change per unit amplitude (nA/nm). If the optical response of the system is known values of photocurrent change can be translated to real height information. Neglecting influences by lateral scanning motion the derived data is equal to the slope of the contact part in a force-distance curve (Fig. 7-11, labelled c).
Compared to a hard sample area the elastic deformation of a soft sample area during force modulation results in a smaller slope in the force distance curve and thus in a lower amplitude in photocurrent difference. Thus SFS is useful in the analysis of topography images in providing an enhanced contrast on surfaces with hardness changes. The greyscale represents the hardness of the sample with low greyvalues being assigned to soft sample areas. As an example the sample area exhibiting the dendritic plate structure discussed previously is displayed in Fig. 7-13. Again the plates reveal their different nature as they appear relatively dark compared to surrounding silicon surface areas. The greyvalue contrast between the plates and silicon can be assigned to the elastic deformation of the surfactant molecules, as discussed as initial deformation following the jump to contact in the force distance curves. It should be noted that for a thorough discussion of elastic properties force distance curves reveal more information about a sample than SFS due to their wider range of applied forces as well as their independence on scanning motion. Nevertheless SFS is useful in providing contrast for surfaces with material changes simultaneously to topography imaging. Furthermore it can be clearly seen in Fig. 7-13 that SFS gives an enhanced contrast to sharp edges such as the plate boundaries or the edges of smaller areas of surfactant molecules ontop of the silicon surface.

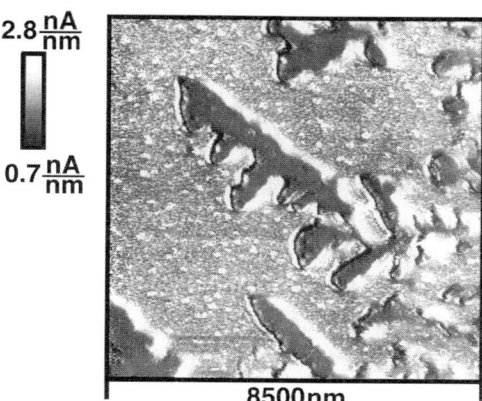

Fig. 7-13: SFS image of the dendritic plate structure shown in Fig. 7-7. Dark greyvalues represent a relatively small response, while bright greyvalues are assigned to a large response to the oscillating sample. The contrast between the dendritic plates and the surrounding silicon surface reveals the higher elasticity of the plates compared with the relatively hard silicon surface.

7.9 Non-Contact AFM

So far contact operation AFM and two of its subdivisions, LFM and SFS have been discussed. All these techniques were operating with a tip being in immediate contact with a surface thus being sensitive to interatomic repulsive forces. As these forces are decaying rapidly with distance van der Waals forces and several others discussed previously in this article become dominant with increasing tip to sample spacing. Thus at tip to sample distance above 1nm a tip experiences mainly these weaker attractive forces. Unfortunately these non-contact regions above the sample surface are not accessible with the contact AFM technique discussed so far. As explained within the discussion of force-distance curves the tip is rapidly pulled onto the surface once the contamination layers overlap or - in vacuum or liquid operation - the van der Waals forces become strong enough. This behaviour resulted from the extremely small spring constants of cantilever systems needed for non-destructive imaging in contact mode.

Thus to utilise van der Waals forces and other attractive far ranging forces for imaging different tip-cantilever systems and imaging techniques have to be considered. For non-contact imaging commercially available cantilevers have spring constants of about 50 nN/nm. Typical capillary forces with values around several 10 nN result in a

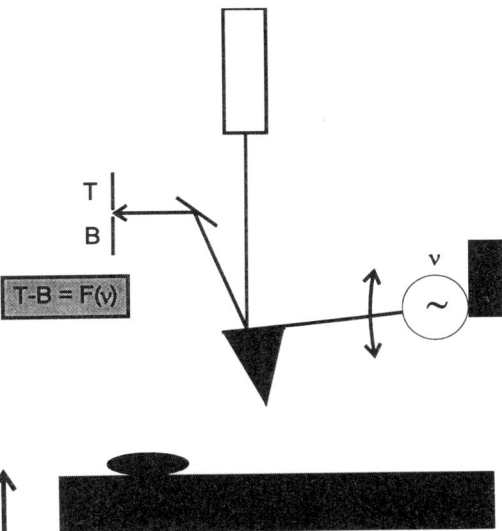

Fig. 7-14: Principle of non-contact AFM set-up. A bimorph stimulates the oscillation of the tip-cantilever system at their resonance frequency. The amplitude of the cantilever oscillation is measured by monitoring the amplitude of the photocurrent difference at the photodiode.

cantilever bending of only a few tenth of a nanometer and are not able to pull the tip onto the surface any more. Of course it is not possible to image a surface with the same operational mode described for contact mode imaging as such a stiff cantilever responds very weakly to any change in force between tip and sample, therefore destroying the sample surface.

Instead the physical phenomena of resonance is used for imaging. For non-contact operation the technical set-up used for contact imaging has to be slightly modified. The tip cantilever system is now attached to a piezoelectric ceramic - a so called bimorph - which is able to oscillate at a wide frequency range (see Fig. 7-14). This frequency range covers the resonance frequency v_{res} of the tip-cantilever system. The relationship between input oscillation and cantilever response is displayed in Fig. 7-15. At frequencies below v_{res} tip and cantilever simply follow the oscillation of the bimorph. As soon as the frequency approaches v_{res} the amplitude of tip-cantilever oscillation increases rapidly. At frequencies above v_{res} the amplitude of the system decreases to low levels again. The value of the resonance frequency itself depends on the spring constant k of the cantilever as well as on the surrounding media. For example a tip-cantilever system will have different resonance frequencies for vacuum, gas or liquid environments. Typical geometries and spring constants used in non-contact imaging yield resonance frequencies ranging from 80 to 300 kHz.

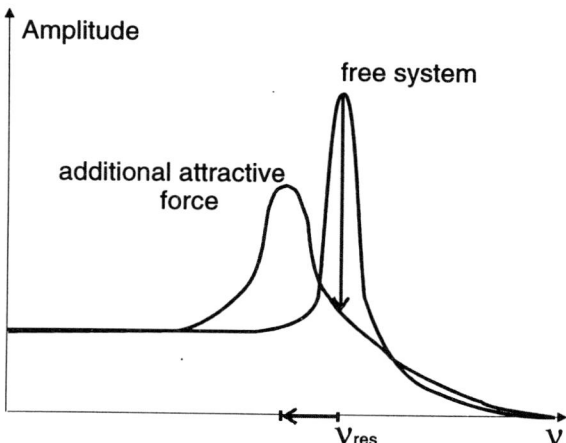

Fig. 7-15: Amplitude to frequency dependence of a free and externally excited system. The external excitation, such as attractive van der Waals forces, causes a shift of the resonance frequency towards lower values. Thus a tip-cantilever system which is excited at its 'free' resonance frequency will decrease its responding amplitude upon approach to a sample surface.

Close to a surface the resonance condition changes if the tip oscillates within the field of an attractive forces above the surface. As the force values decrease with distance they will influence the tip inhomogeneously, i.e. they will be larger at the part close to the surface during oscillating motion compared to the more remote part of it. This change in force - the so called force gradient $\partial F/\partial d$ - enhances oscillation for attractive forces and causes a shift of the resonance amplitude towards lower values (see Fig. 7-15). Mathematically the system can be treated as if the spring constant k of the tip cantilever system has changed to

$$k' = k - \frac{\partial F}{\partial d}.$$

For AFM operation one has to monitor a quantity which depends on the distance between tip and surface. In an ambient air environment the first overlap of the contamination layers causes a large force gradient. Nevertheless once the tip is oscillating within the contamination layer of the sample surface the influence of the capillary forces can be rather small as they do not change then as strongly with distance to the surface as the other attractive forces do. In the equation given to calculate the strength of capillary forces (see above) the diameter of the water bridge 2r between tip and sample does not change for small changes in d. Consequently the derivative $\partial F/\partial d$ for the capillary force yields a constant value, i.e. its influence on tip oscillation independent of the distance between tip and surface. In contrast van der

Waals forces decrease with $1/d^2$ and thus yield comparatively large changes in force gradients with distance. Therefore the influence of the contamination layer can be neglected so that this modulation technique depends mostly on van der Waals forces.

Besides the use of different cantilevers a further change in the technical set-up is applied to the control electronics in non-contact operation AFM. Now the signal from the photodiode is processed by a lock-in amplifier. Only signal components matching the initial oscillation exciting the tip-cantilever system are amplified while other components are discriminated. The output voltage is proportional to the oscillation amplitude of the matching signal components. Thus by acquiring the output voltage of the amplifier the tip amplitude to a given frequency can be measured.

In a typical mode of operation the system is tuned with the tip far above the surface so that the tip-cantilever system is excited at about its resonance frequency v_{res}. As the tip approaches the surface van der Waals forces will change the resonance properties. This is monitored as a decrease in oscillation amplitude (marked by the vertical arrow in Fig. 7-15) for the tip-cantilever system is still excited at the former resonance frequency v_{res}. Thus the approach can be stopped at a given damping value of the amplitude, which serves as setpoint in non-contact operation as compared to the difference in photocurrent value chosen in contact operation AFM. During scanning motion a protrusion in topography will further reduce the amplitude, while a depression will cause an increase. The system is able to respond to these topography changes if the amplitude serves as signal for the feedback loop.

As van der Waals forces are dominant in non-contact operation the exchange forces acting between tip and sample can be kept to values which are two to three orders of magnitude lower than during contact operation. This allows imaging of delicate samples, which would be either destroyed or otherwise influenced during contact imaging. For example Fig. 7-16 displays drug loaded SLN. For contact mode imaging the particles had to be placed on an especially prepared silicon surface. On one hand the surface had to be smooth enough to guarantee an unambiguous distinction between surface topography features and the particles. On the other hand the surface roughness had to be increased to avoid that the particles were simply pushed out of the imaging area by the tip. In contrast for non-contact mode operation it was sufficient to place the particles on a glass slide thus simplifying sample preparation. With the low exchange force it can be guaranteed that the particle shape itself is not influenced by tip to sample forces. It should be noted in this context that contact and non-contact imaging of the particles yielded the same particle sizes, which agree well with those measured by photocorrelation spectroscopy (PCS).

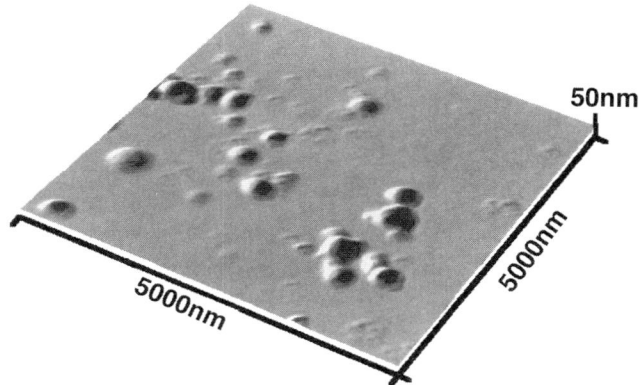

Fig. 7-16: Non-contact AFM image of SLN. The scanarea covers a 5000nm x 5000nm large sample area. The image is represented in a so called 'shaded' display. Greyvalues represent the slope between neighbouring data points towards the upper left of the image, giving the impression of a light source being placed there. In addition the image is displayed three dimensionally. About a dozen SLN can be identified in the image (from [15]).

7.10 Conclusion

AFM and STM are both able to image nanostructures in three dimensions with resolutions down to the atomic scale. Besides pure topography imaging AFM is capable of acquiring tribological and elasticity information. With their ability to operate in air, liquids and under vacuum conditions AFM and STM offer a wide range of possible applications. The relative ease of sample preparation overcomes problems arising in electron microscopy. It should be noted that quite often electron microscopy and SPM techniques offer access to complementary information. For example SPM techniques find best imaging conditions on very flat or insulating samples, overcoming resolution limits of SEM. In contrast very rough samples can cause a large amount of tip artefacts in an image due to its geometrical properties suggesting the preferred use of an SEM. With the ability to image in transmission TEM offers insight into a thin sample structure while standard SPM techniques are limited to surface information.

7.11 References

[1] Bergmann Schäfer, Lehrbuch der Experimentalphysik, Bd. 3 Optik, p 406 ff.

[2] S.L. Flegler, J.W. Heckman, K.L. Klomparens, Elektronenmikroskopie, Spektrum Akademischer Verlag, 1995

[3] R. Wiesendanger, Scanning Probe Microscopy and Spectroscopy, Cambridge University Press (1994)

[4] G. Binnig, H. Rohrer, Helv. Phys. Acta 55, 726 (1982)

[5] J. Tersoff, D.R: Hamann, Phys. Rev. B 31, 805 (1985)

[6] H. Niehus, Appl. Phys. A 53, 388 (1991)

[7] M. Amrein, R. Durr, A. Staslak, H. Gross, G. Tavaglini, Science 243, 1708 (1988)

[8] P.G. Arscott, G. Lee, V.A. Bloomfield, D.F. Evans, Nature 339, 484 (1989)

[9] U. Hartmann in Springer Series in Surface Science Vol. 29; Scanning Tunneling Microscopy III; Eds.: R. Wiesendanger, H.-J. Güntherodt; Springer, Berlin, Heidelberg 1993, pp. 293-359

[10] P.W. Atkins, Physical Chemistry, 3rd Ed., Oxford University Press

[11] G. Binnig, C.F. Quate, Ch. Gerber, Phys. Rev. Lett. 56, 930 (1986)

[12] O. Marti, B. Drake, P.K. Hansma, Appl. Phys. Lett. 51 (7), 484 (1987)

[13] Y. Martin, C.C. Williams, H.K. Wickramasinghe, J. Apl. Phys. 61, 4723 (1987)

[14] G. Meyer, N.M. Amer, Appl. Phys. Lett. 53, 1045 (1988)

[15] A. zur Mühlen, E. zur Mühlen, H. Niehus and W. Mehnert, Pharm. Res. 13 (9), 1411 (1996)

[16] M. Nonnenmacher, doctoral thesis p. 38, Fachbereich Physik, Gesamthochschule Kassel

[17] 'Nanotribology', MRS Bulletin Volume 18 (5), (1993)

[18] P. Jacob, Y.J. Chabal, J. Chem. Phys. 95, 2897 (1991)

[19] Z. Shao, J. Yang, Quarterly Rewiew of Biophysics 28 (2), 195 (1995)

[20] E. Henderson, Progress in Surface Science 46 (1), 39 (1994)

[21] Topometrix TMX 2000, Topometrix Corp., Santa Clara, CA, USA

Address of the authors:
Dr. Ekkehard zur Mühlen, Prof. Dr. Horst Niehus
Institut für Physik/ASP
Invalidenstr. 110
D-10115 Berlin
Germany

8 FTIR Techniques for Surface Characterization

Dr. A. Büchtemann, Fraunhofer Institute, Teltow and Dr. R. Dietel, University Potsdam

8.1 Introduction

The subject of spectroscopy is the investigation of the interaction of electromagnetic radiation and matter, as related to the absorption and/or emission of radiation of definite wavelength. The infrared spectroscopy covers that segment of the electromagnetic spectrum that lies between the longwave limit of visible light (about 0,8µm) and the longwave limit of thermic sources of radiation (about 1000µm)(Fig. 8-1).

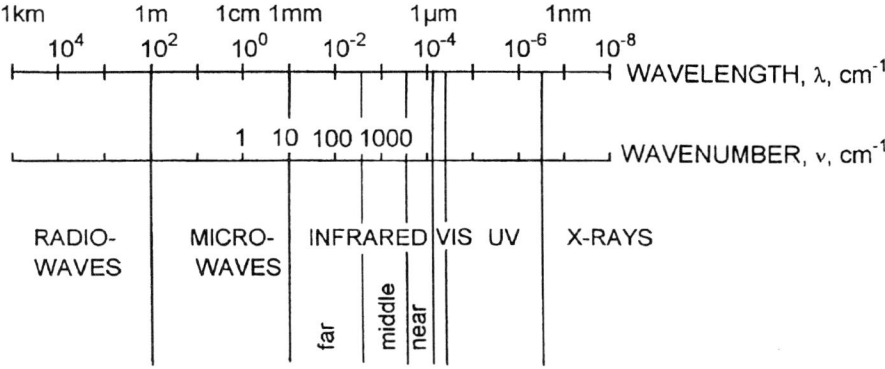

Fig. 8-1: Part of the electromagnetic spectrum demonstrating wavelengths and wavenumbers of the infrared region

The infrared spectroscopy is a versatile and effective analytical method [1]. It produces definite results of the chemical constitition of investigated compounds, it can be used for qualitative and for quantitative analysis and provides physical information of molecular structure and of specific properties of substances. These results range from information on the molecular conformation to the orientation of characteristic groups in the molecules.

Today a large number of techniques for infrared spectroscopic research is available for many different purposes [2]. They differ by their specific fields of application, the demands on the preparation (that is the manner and form in which the object or substance should be present), the specific available information and the manner in which the measurement reacts to the refractive index and the absorption index.

There are several techniques available for infrared spectroscopic investigations which exclude physical contact and damaging effects; bulk properties can be studied as well as the properties of surfaces, and with the use of infrared microscopy even spectra of microscopic objects or details thereof can be recorded [3,4].

Infrared spectroscopic investigations can be performed in transmission and in reflection, within the limitations given by the particular object studied. The techniques, especially the reflection technique, can be subdivided into several, quite different specific procedures. The basis for and the specific qualities of some of the techniques which are of special importance for the infrared spectroscopic characterization of surfaces, interfaces and thin films will be shown in the following.

8.2 Basic Information on Infrared Spectroscopy

Infrared radiation can be described as electromagnetic wave of the wavelength λ, the frequency ν or - as commonly called within the infrared spectroscopy - the wavenumber $\overline{\nu}$, respectively. These quantities are related as follows:

$$\nu = c/\lambda; \; \overline{\nu} = 1/\lambda; \; \overline{\nu} = \nu/c \qquad \text{Eq. 8-1}$$

(c - velocity of light). The energy of infrared electromagnetic waves is directly proportional to the frequency (or wavenumber):

$$E = h \cdot \nu = h \cdot c/\lambda = h \cdot c \cdot \overline{\nu} \qquad \text{Eq. 8-2}$$

($h = 6,625 \cdot 10^{-34}$ Ws^2, Planck's constant).

The infrared spectroscopy comprises the range of energy of vibrations and rotations of molecules and/or segments of molecules, registering overtone bands and combination vibrations in the near infrared (0.8µm - 2.5µm; 12500cm^{-1} - 4000cm^{-1}), basic vibrations (valence and deformation oscillations) in the middle infrared (2.5µm - 25µm; 4000cm^{-1} - 400cm^{-1}), and skeleton (e.g. with polymers) or lattice vibrations (e.g. in crystals) in the far infrared (25µm - about 1000µm; 400cm^{-1} - 10cm^{-1}). Fig. 8-2 shows schematically the fundamental vibrations of a three-atomic, non-linear molecule (e.g. water); they belong to the middle infrared spectral region which this contribution is confined to in the following.

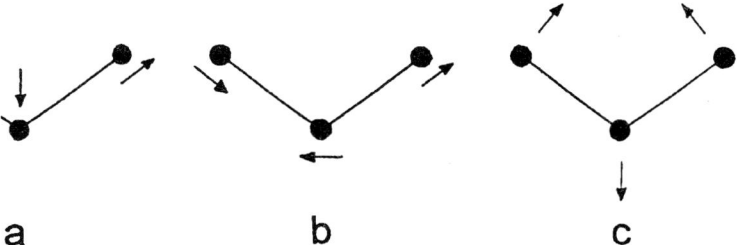

Fig. 8-2: Fundamental vibrations of a three-atomic, non-linear molecule; a - symmetric stretching vibration, b - asymmetric stretching vibration, c - deformation vibration

Motions of vibration or rotation of atoms within molecular structure can be described by the model of the anharmonic oscillator or the non-rigid rotator, respectively [5]. A system of discrete, non-equidistant energy values (terms) results for the energy of these motions.

Elementary interaction between infrared radiation and matter, that is to say the absorption of radiation energy by molecules and the accompanying stimulation of movements of vibration and rotation can ensue only in discrete energy quanta. This is expressed by the so-called frequency and resonance condition of Eq. 8-3.

$$E'' - E' = h \cdot \nu = h \cdot c \cdot \overline{\nu} \qquad (E'' > E') \qquad \text{Eq. 8-3}$$

According to this, radiation is only absorbed with such energy that corresponds to the difference between two discrete energy levels within the term scheme of the molecule. The wavenumber $\overline{\nu}$ of the absorbed radiation results directly from the term difference (E''-E').

A further condition for the absorption of infrared radiation is the necessity that the vibration to be stimulated must be accompanied by a change of the dipole moment. Change of dipole moment results from the motion of the charge center if the molecule (or the corresponding atomic grouping) contains a permanent asymmetrical charge distribution or if the latter is induced by the motion of oscillation.

Due to the characteristic ensemble of motions of the vibration and rotation of every molecule the absorption of infrared radiation is by itself a specific property of the substance; the measuring and interpretation of these motions is the subject of infrared spectroscopy.

The interpretation of infrared spectra can be performed in different ways:

A precise interpretation can be achieved by the normal coordinate analysis: every vibration is described by simply superimposing the normal vibrations of a given molecule. (When dealing with macromolecules the analysis of the normal coordinates is based on the study of the possible vibrations in the repeating unit [6,7]). The frequencies of normal vibrations (3N-6 for non-linear and 3N-5 for linear molecules, N being the number of atoms within the molecule) for simple molecules (few atoms) can be calculated precisely from the equations of motion of all the atoms of the molecule if the molecular data (atomic masses, bond lengths, force constants of the bonds) are available. Every absorption band in the IR-spectrum can then be assigned to a characteristic vibration of the molecule considered. More complicated molecules (many different atoms, single and multiple bonds) demand - despite the application of a variety of approximation and iteration procedures - much more time for the normal coordinate analysis. It is therefore only suited for select cases interpreting IR-spectra of the routine analytical work. However, because of its precise description of the vibrations it is of great importance for the determination of the moments of inertia, bond lengths, force constants and symmetry properties.

For practical purposes, the phenomenological model of the characteristic group frequencies is a favoured concept for the interpretation of infrared spectra [5,8,9]. This concept is based on the presumption that due to the diversity and discontinuity of distribution of the atomic masses and bond forces the molecule contains oscillators with differing frequencies; the more the oscillation energies differ, the less they influence each other. The corresponding frequencies of vibrations can be assigned to specific atomic groupings in the molecule; thus they are called characteristic group frequencies. Their assignment follows empirical rules with the assistance of the normal coordinate analysis. Since the motion of vibration of one group is only in approximation not influenced by the motions of neighbouring groups, the characteristic group frequencies are not definite numbers, but they comprise certain frequency ranges reflecting the neighbourhood of the atomic group considered (Fig. 8-3).

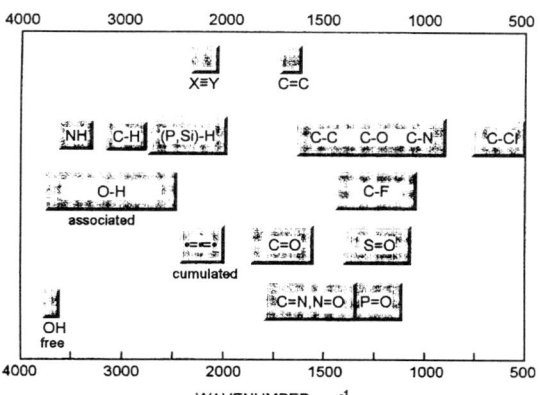

Fig. 8-3: Scheme of some characteristic group frequencies

Thus, very specific spectra interpretation is possible. The ranges of characteristic frequencies overlap to certain extent. In some cases this fact can considerably impede the interpretation of the spectra by use of characteristic frequencies. Therefore spectra of complicated compounds may be interpreted successfully by a comparison with spectra of authentic substances.

Hence, the comparison of spectra is another possibility of the interpretation of IR-spectra. It is supported by spectra collections which are arranged to substance classes [10,11,12]. For the use of modern, computerized spectrometers electronic libraries are available. Within a minimum of time they provide a comparison of the recorded spectra with a large number of catalogued spectra. This computer-based search is an effective tool for the interpretation of spectra; however, in certain cases, especially when analysing mixtures and very special or new substances its usefulness is limited. In practice, the interpretation according to the concept of characteristic group frequencies is often combined with the comparison of spectra.

8.3 Transmission Spectroscopy

Transmission measurements, that is measuring the portion of the incident electromagnetic energy (Fig. 8-4) which passes through an object, basically are not predestined for infrared spectroscopic investigation of boundary layers, near-surface layers or even surfaces themselves since it is aimed at inspecting exactly the energy absorbed in the <u>volume</u>.

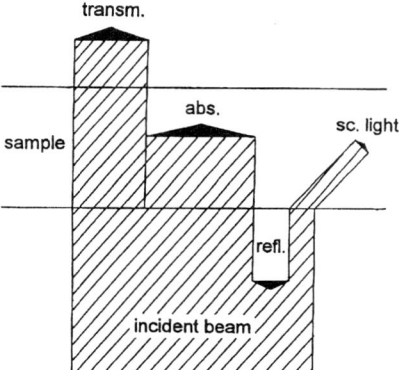

Fig. 8-4: Intensity components of a light beam interacting with a sample; transm. - transmitted light, abs. - absorbed light, refl. - reflected light, sc. light - scattered light

Today, however, modern IR-spectrometers, especially when employing special procedures of measurement and evaluation, make possible the recording of transmission spectra of very small amounts of substances or very thin layers with a tickness of far below 1μm [13,14]. Transmission measurements supply in many cases a spectrum that depicts quite well the dependence of the energy actually absorbed by the studied object on the wavelength. That is the reason why transmission spectra so often are of utmost importance; they should be chosen for purposes of comparison even if, in a concrete case, they were recorded from a large amount of substance or a thicker layer than the investigated object can offer.

The intensity of a monochromatic electromagnetic light wave which is propagating in the absorbing medium, decreases following the law of absorption

$$J = J_0 \cdot e^{-\alpha' X} \text{ or } J = J_0 \cdot 10^{-\alpha X} \qquad \text{Eq. 8-4}$$

In this relation, commonly known as **Lambert's Law**, J_0 and J signify the radiation intensity at the beginning and the end of a distance of the length of X, travelled by the wave in the medium, and α' and α the natural and decadic absorption coefficient, respectively, of the medium for this wavelength. The product $\alpha \cdot X$ is called **absorbance**. The size of the absorption coefficient α and its spectral course is determined by the molecular chemical and physical structure of the medium as well as by the number of molecules or molecular segments which can be stimulated to oscillate in the penetrated volume, that is, their concentration. If this number remains constant then the concentration need not explicitely appear in Eq. 8-4. In numerous cases however (e.g. dispersions, solutions, mixtures) the number might vary: we find here, under the condition of a linear connection between absorbance and concentration, the well-known **Lambert-Beer's law** (5)

$$J = J_0 \cdot 10^{-\varepsilon c X} \qquad \text{Eq. 8-5}$$

Describing the concentration in mol/l and the distance X in cm the physical quantity ε, called molar extinction coefficient (or absorptivity /1/), has the dimension $l \cdot mol^{-1} \cdot cm^{-1}$. As commonly done today [5,6,7], we used the number 10 as the basis for the power in Eq. 8-5. If e is to be used as the basis the values of absorbance and absorption coefficient will increase by the factor $\ln 10 = 2,303...$; attention to this fact must be paid when comparing with texts.

The relationships between α (or ε, respectively) and the quantities k (absorption index) and κ (attenuation index [20]), used in the field of physical optics as well as in infrared spectroscopy, which are defined by the (complex) refractive index

$$\tilde{n} = n + ik = n(1 + i\kappa) \qquad \text{Eq. 8-6}$$

are given by

$$\alpha = (4\pi/\ln 10)\nu \cdot k = (4\pi/\ln 10)\nu \cdot n \cdot \kappa \qquad \text{Eq. 8-7}$$

and

$$k = (\varepsilon \cdot c \cdot \ln 10)/(4\pi \cdot \nu) \qquad \text{Eq. 8-8}$$

Unfortunately, the terms used in literature are not standardized; we follow here the books of BORN/WOLF [20] and HENNIKER [15].

Lambert-Beer's law is the basis for the quantitative evaluation of the infrared spectra measured in transmission. Some remarks concerning its application may be of use, especially for the investigation of thin layers.

First it has to be pointed out that the relationship of absorbance and concentration is linear only if the concentration is not too high (e.g. no associate formation); if necessary, linearity must be proved by measuring differing concentrations.

When studying thin layers attention must be paid to the fact that Eq. 8-5 describes the decrease of intensity due to absorption on the distance X exclusively in a medium being homogeneous concerning the refractive index (that is, there are no changes in the refractive index). However, when penetrating an object additional intensity losses by reflection on its surface, possibly even by scattering on structures of the surface or inhomogeneities take place (Fig. 8-4). When working with flat objects that show hardly any deviation in thickness and have plane surfaces, as with thin films, often superposition of the spectrum with a wavelike basis line (interferences) occurs due to multiple reflections on the surfaces. The **transmittance T** measured experimentally on the object, that is, the ratio of the transmitted to the incident intensity, is, for the given reasons, not identical with the quotient J/J_0 as defined by Lambert-Beer's law, and the transmission spectrum (be it shown in transmittance T or absorbance A) does not correspond to the true course of α. The changes in the transmission spectrum due to reflection play an important part especially with thin objects since these changes depend solely on the refractive indices of object and environment and not on the thickness of the layer. They gain therefore with decreasing thickness of the layer - and the accompanying decreasing absorption - in relevance so that finally even a loss of intensity of several percent, due to reflection, cannot be neglected.

Despite these influences transmission spectra can be evaluated in most cases according to the base line procedure (Fig. 8-5) [5,6,7].

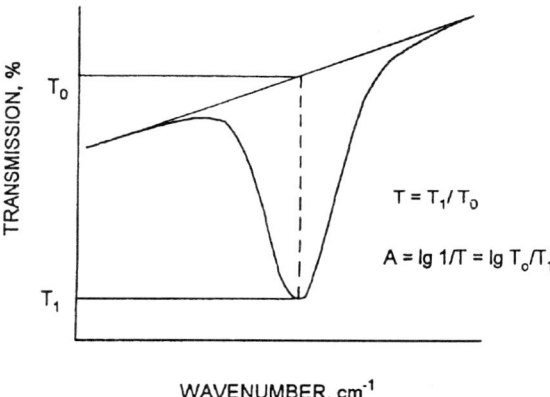

Fig. 8-5: Scheme of base-line method; T - transmittance, A - absorbance

The reason for that is that the above mentioned effects usually do not depend strongly on wavelength, that is, their influence at the maximum of an absorption band and its neighbourhood is of similar magnitude so that they, when relating the bands to their environment, will be mostly eliminated. This is true e.g. for the losses from scattering which can generally be noticed in a more or less major decrease of transmittance towards the shortwave region of the spectrum as well as for the interferences whose period increases with very thin objects, so that the baseline procedure can well be applied. The remaining changes of band intensities and shapes due to reflection, especially when accompanied by strong absorptions (large k,α,ε) are generally small in polymers. In consideration of these changes special evaluation algorithms and, if necessary, simulation calculations need to be applied [16,17,18].

The preparation of the substances for the recording of a transmission spectrum depends on the condition of the original material and the purpose of the investigation. The KBr press technique often provides good results for solids; for solutions and dispersions good results are obtained by studying in a capillary layer or in cuvettes or by the preparation of films of controllable thickness on transparent substrates (e.g. by dipping, casting or spin-coating) [19]. When preparing layers for transmission studies locally varying thicknesses are to be avoided; otherwise the band intensities in the spectrum will be falsified. Details of preparation methods can be found e.g. in [5,6,7,15].

Fig. 8-6 shows an example of a transmission spectrum of a disperse system (The latices used as examples in this contribution were synthesized by Dr. B.-R. Paulke, Fraunhofer Institute of Applied Polymer Research.).

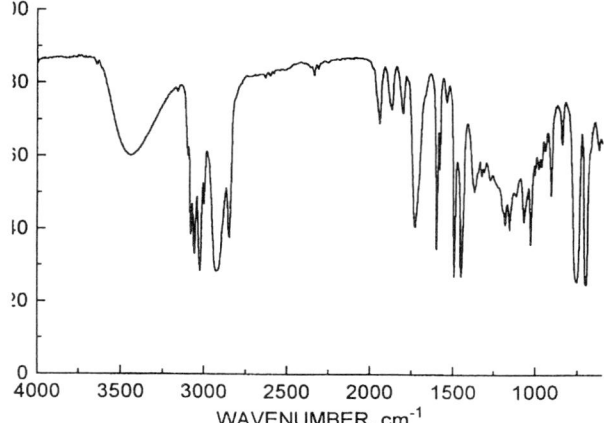

Fig. 8-6: Transmission spectrum of a dried layer of a core-shell-latex (same latex as in Fig. 8-11, Fig. 8-14c, and Fig. 8-18); core: polystyrene, shell: poly-2,3-epoxypropyl-methacrylate

The system is a core-shell-latex consisting of a polystyrene core and an epoxyacrylate shell. Preparation by drying resulted in a brittle, cracked layer; thus a certain portion of light not passing through the substance, but contributing to the spectrum could not be avoided. The spectrum shows clearly bands of polystyrene (e.g. 1600cm^{-1}, 1585cm^{-1}, 1500cm^{-1}, 1460cm^{-1}) as well as bands due to acrylate (1720cm^{-1}). The broad band at about 3450cm^{-1} indicates OH groups formed by hydrolysis of the oxirane ring. The assumed particle shape is shown in Fig. 8-7.

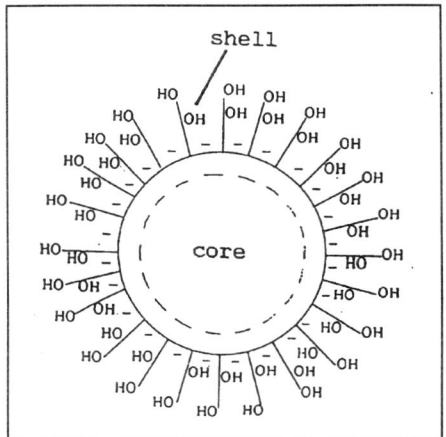

Fig. 8-7: Assumed particle shape of the latex shown in Fig. 8-6

8.4 Reflection Spectroscopy

8.4.1 External Reflection

The external reflection spectroscopy is an especially effective technique for the investigation of surfaces and thin layers. However, the shape of measured reflection spectra is different from the shape of spectra usually found in transmission spectroscopy. A number of factors are responsible for that and have to be considered for a successful interpretation of external reflection spectra. Therefore some important phenomena will be briefly explained here.

If infrared radiation strikes the plane interface between two optical media (e.g. the surface of a solid) part of the radiation will be reflected. Another part penetrates the interface and propagates beyond it (inside the solid, Fig. 8-8).

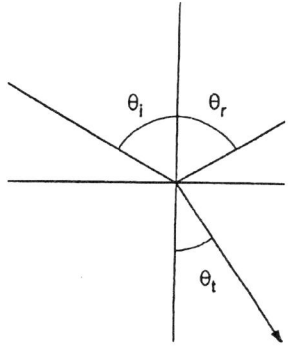

Fig. 8-8: Reflection and refraction of light at a plane boundary,
q_i - angle of incidence,
q_r - angle of reflection,
q_t - angle of refraction;
plane of incidence corresponds to paper plane

The specularly reflected part of the radiation is recorded by the external reflection spectroscopy. It measures the degree of reflection of this part of the radiation, that is the ratio of reflected and incident radiation intensity, in dependence on the wavelength. In actual practice, the incident intensity is hardly ever measured directly. The measured intensity from a good reflection mirror takes its place, and is then put in relation to the reflected intensity of the sample.

The propagation of infrared radiation striking the plane interface between two optical media can be described completely by the law of reflection, the law of refraction, and Fresnel's equations [20]. The law of reflection

$$\theta_i = \theta_r \qquad \text{Eq. 8-9}$$

(θ_i: angle of incidence, θ_r: angle of reflection) and the law of refraction

$$n_1 \cdot \sin\theta_i = n_2 \cdot \sin\theta_t \qquad \text{Eq. 8-10}$$

(θ_t: angle of refraction, n_1: refractive index of the medium in which the incident and reflected rays are propagating, n_2: refractive index of the medium in which the refracted rays are propagating) describe the **direction** of the reflected and the refracted rays. They are located, together with the incident radiation and the normal to the interface, in one plane, the so-called plane of incidence (see Fig. 8-8).
Fresnel's equations, however,

$$r_s = (n_1 \cos\theta_i - n_2 \cos\theta_t)/(n_1 \cos\theta_i + n_2 \cos\theta_t) \qquad \text{Eq. 8-11}$$

$$r_p = (n_2 \cos\theta_i - n_1 \cos\theta_t)/(n_2 \cos\theta_i + n_1 \cos\theta_t) \qquad \text{Eq. 8-12}$$

describe the **intensity** of the reflected light (r_S or r_P: relative amplitudes of the reflected wave for the polarization directions perpendicular or parallel to the plane of incidence). It is determined by the angle of incidence, by the state of polarization of the incident light and especially by the optical properties of the two media that are separated by the interface layer. In the experiment, the angle of incidence and the state of polarization are set by the reflection equipment and the polarizer used. For quantitative reflection spectroscopic studies working with polarized radiation is absolutely necessary, since, according to Eq. 8-11 and Eq. 8-12, r_S and r_P have different values. The optical properties of the media involved are described in Fresnel's equations (Eq. 8-11, Eq. 8-12) by the corresponding refractive indices. For absorbing media, the refractive index following Eq. (6) is always a complex quantity. While according to Lambert-Beer's law and Eq. 8-7 and Eq. 8-8 only the absorption index k (or the attenuation index κ, resp.) of the investigated medium is of importance for the absorption in the bulk (that is in case of transmission measurements), Fresnel's equations show that the intensity of reflected radiation in addition to the dependency on k is <u>also</u> dependent on the real part n of the refractive indices of the media involved. This major difference between transmission and reflection experiments leads to the changed shape of reflection spectra which is often found in comparison to transmission spectra. This has to be taken into account when interpreting reflection spectra [21].

The experimental setup determines how the above mentioned correlations influence the recording of reflection spectra. If a sample consisting of several absorbing layers (optical media) is to be investigated in the reflection experiment, the law of reflection and the refraction law as well as Eq. 8-11 and Eq. 8-12 will have to be observed for every interface; additionally Lambert-Beer's law has to be taken into account for the propagation of the IR-radiation in the bulk between the interfaces [20,22]. In actual practice IR-radiation, in general, propagates through air as the first medium (that is $n_1=1$)

and hits not more than two boundaries, so that the consideration is reduced to maximally three different media [23].

Most of the experiments can be assigned to one of the following cases which have to be treated differently according to the optical properties of the investigated samples:

a) reflection at the surface of a thick (theoretically "half-infinite") sample (solid or liquid), that is **one** boundary layer (see Fig. 8-8),
b) reflection at a free standing film (**two** boundary layers),
c) reflection at a covered metal surface (**two** boundary layers),
d) reflection at a covered surface of a thick (half-infinite) dielectric or semiconductor (**two** boundary layers).

Fig. 8-9 shows the dependence of the intensity of radiation reflected at the surface of a dielectric or semicinductor on the angle of incidence. The radiation used is polarized either parallel (p,II) or perpendicular (s,⊥) to the plane of incidence.

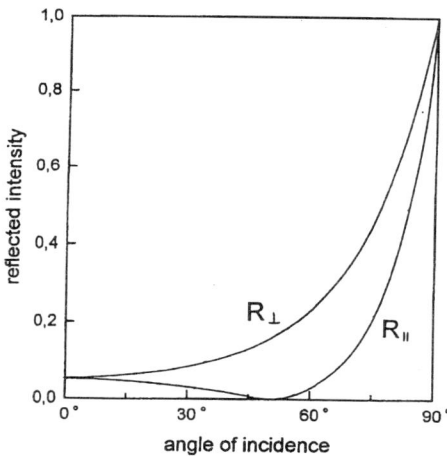

Fig. 8-9: Reflection degree R in dependence on angle of incidence and polarization direction in case of reflection at the surface of a dielectrics or semi-conductor

The different course for the two states of polarization can be easily recognized. While the reflected intensity for s-polarization increases steadily with the increasing angle of incidence, the p-polarized radiation shows a minimum. The angle of incidence related to this minimum is the so-called **Brewster angle** θ_B which is dependent on the refractive index of the sample. θ_B plays an important part in the field of reflection spectroscopy and must be considered especially when examining absorbing layers on the surface of dielectrics or semiconductors (case d) [20,24]. Here it should only be mentioned that the intensities of the bands of a reflection spectrum depend (among others) on the relation between the angle of incidence θ_i and θ_B. For a coated dielectric the quotient θ_i/θ_B (larger or smaller than 1) determines, whether a band in the reflection spectrum is directed "upwards" or "downwards", respectively [25,26].

The intensity of p-polarized radiation when reflected at metal surfaces does not show a minimum as we find in Fig. 8-9. Ultimately, the different optical properties and conductivities of metals and non-metals are the reason for this different behaviour as well as for the very large Brewster angles of metals (near 90°, measured to the surface normal) [27].

Using examples a) and c) we will demonstrate the different optical behaviour of metal and non-metal surfaces in reflection experiments. In actual practice these examples are the most numerous and show the most important differences between transmission and reflection spectroscopy.

The behaviour of the intensity of the reflected radiation in dependence on the angle of incidence which we dealt with here, is quite important for the application of the external reflection spectroscopy when examining the surface of compact materials (case a). Here we have dielectrics and semiconductors first and foremost in mind. From such experiments the two components of the complex refractive index, that is n **and** k, in its dependence on wavelength can be determined for the studied material or its surface. Due to their dispersion-like character (i.e. they reproduce to some extent the dependence of n on the wavelength) the reflection spectra recorded at the surface of a compact material show only minimal similarities to the transmission spectra of the very same material (Fig. 8-10).

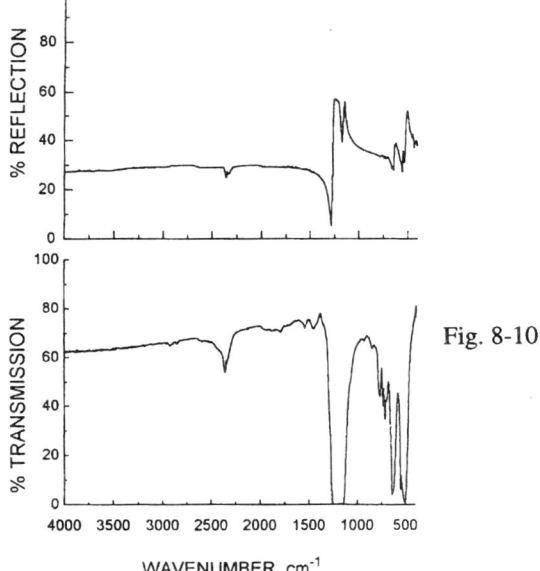

Fig. 8-10: Comparison of the reflection spectrum of polytetra-fluoro-ethylene (angle of incidence 80°, s-polarization, sample thickness 2 mm) with a transmission spectrum of the material (cold-pressed sample)

However, qualitative findings can be obtained up to a certain degree, by an experienced spectroscopist, directly from the measured reflection spectra [21].

A valuable tool for a quantitative interpretation is the Kramers-Kronig transformation. By means of this procedure the measured reflection spectra can be transformed into a shape as in transmission spectroscopy and the real as well as the imaginary part of the refractive index can be calculated [18]. The commercially available software for the evaluation of reflection spectra often contains algorithms for the Kramers-Kronig transformation and for spectra simulations. The approximation of the calculated and measured spectra permits - within the framework of certain models - the quantitative determination of n and k or the parameters which are responsible for the optical properties of the studied material. Often the combination of Kramers-Kronig analysis and spectra simulation makes sense or is even necessary.

If absorbing layers on metal surfaces are investigated (case c) recorded reflection spectra display - in contrast to case a - a shape similar to the transmission spectra, that means the shape of the spectra is mainly influenced by the absorption properties of the studied layers. The metal surface operates merely as a reflecting substrate and its optical parameters do not influence the recorded spectra noticeably. This statement is valid for "thick" layers ($d > \lambda$, d:film thickness). The investigation of very thin layers ($d \ll \lambda$) is much more complicated. In this case one has to consider that a standing wave is set up by the interaction of incident and reflected radiation at the metal surface, however, for s- and p-polarization each in a specific manner. For s-polarization there appears a knot near the metal surface independent of the angle of incidence, that is, the amplitude of the electric field of the standing wave is very small in the direct vicinity of the surface. In the case of p-polarization the amplitude of the electric field increases with increasing angle of incidence beginning close to zero at perpendicular and reaching a maximum at grazing incidence (angle of incidence near 90°) [2,27].

Fig. 8-11 demonstrates this behaviour very well, by showing different reflection spectra of a thin latex layer prepared upon a gold substrate.

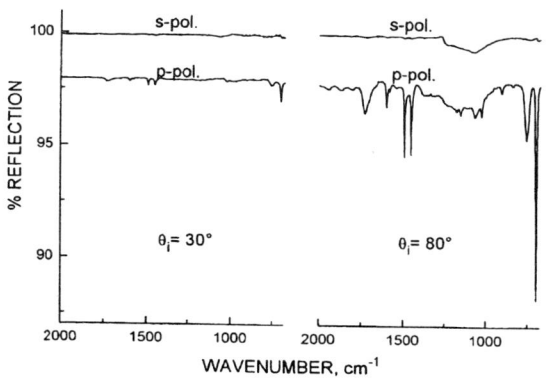

Fig. 8-11: Reflection spectrum of a thin dried latex layer on gold for different angles of incidence (q_i) and polarization directions, same latex as Fig. 8-6.

For s-polarization in the immediate vicinity of the metal surface there is nearly no interaction between the IR-radiation and the latex layer. Thus no absorption bands of the latex layer appear in the recorded spectra neither at small (30°) nor at large angles of incidence (80°). Therefore, the investigation of layers on metal surfaces with s-polarized radiation will only show results when working with thicker layers ($d \geq \lambda$). Contrary to that, at grazing incidence and p-polarized radiation there is an intensive interaction between IR-radiation and molecules which are located directly on the metal surface. Here the absorption bands of the latex layer appear in the specific spectra of Fig. 8-11 and are especially intense at grazing incidence (80°). This behaviour is taken advantage of for the study of extremely thin layers ($d \ll \lambda$, up to monomolecular films) [28]. Attention must be paid to the fact that due to the state of polarization of the incident radiation only vibrations with transition moments perpendicular to the surface will be excited. If anisotropic layers are dealt with this phenomenon can be used for the determination of the molecular orientation of the sample [29].

The external reflection spectroscopy works without contact or destruction. On the one hand it detects the optical properties of compact materials and is on the other hand an especially powerful tool for the examination of thin films. Its application ranges from, among others, the investigation of adsorbed molecules on solid (catalysis) or liquid surfaces (environmental research), to the detection of contaminations on semiconductor materials (microelectronics) and the clarification of adhesion phenomena (research on adhesives). A theoretical basis is needed for the application to a rather large degree. Therefore this analytical technique is employed mostly at universities and other research institutes at present. However, by reason of its special suitability for in-situ-analysis it is also of interest for industrial research.

8.4.2 Attenuated Total Reflection (ATR)

The attenuated total reflection spectroscopy can be viewed as being positioned between transmission and reflection spectroscopy since it analyses the surface as well as a region of variable thickness below the surface as determined by the experimental conditions. Additional terms used for this technique or special variants thereof are "(Multiple) Internal Spectroscopy (IRS)" and "Frustrated Total Reflection (FTR)".

The ATR is a powerful infrared spectroscopic method that has been successfully employed to solve a large number of problems relevant as far as surfaces are concerned [30,31,32].

The result of ATR measurements is determined much more than it is the case in transmission experiments by the choice of the experimental parameters. The correct choice is crucial for the full achievement of the potential of this method. In the following some fundamental concepts of this method are presented; concerning the complicated details, refer to the literature [30,33,34].

When an electromagnetic wave from a denser medium (n_1) strikes the interface to a rarer medium (n_2) under an angle of incidence larger than the critical angle θ_c, given by Eq. 8-13,

$$\sin \theta_c = n_2/n_1 \qquad \text{Eq. 8-13}$$

total reflection takes place (Fig. 8-12). That means the intensities of the incident and reflected rays are identical. Refraction of light into the rarer medium does not occur.

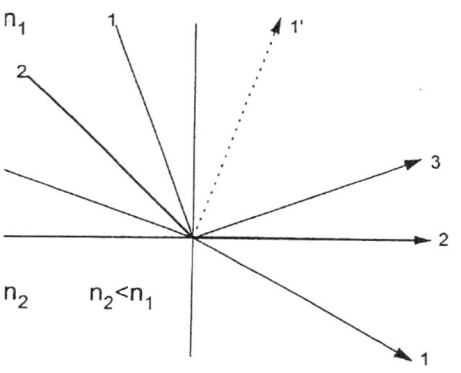

Fig. 8-12: Transition from specular reflection and refraction to total reflection at increasing angles of incidence q_i. The angle of incidence of ray 2 corresponds to the critical angle (q_c) of total reflection.

The medium n_1 which has to possess not only a large refractive index, but also a high transparency in a wide infrared spectral region, is realized by the so-called ATR element (ATR crystal). The rarer medium is the sample which must be in the closest possible contact to the ATR element. In the ATR experiment the IR radiation coming from the light source enters the ATR crystal and interacts with the sample.

Despite of total reflection, certain penetration of light into regions near zo the surface of the rarer medium (i.e. of the sample) occurs during the reflection process. If this medium absorbes light some energy of the reflected radiation is lost within the absorbing spectral regions. Strictly speaking, we therefore do not have total reflection, but <u>attenuated</u> total reflection. Measuring the spectral distribution of intensity of the reflected light provides an absorption spectrum of the sample. However, due to the specifics of its origin this spectrum is, in general, different from the absorption spectrum obtained in transmission spectroscopy [33].

The strength of the electric field created in the rarer medium due to the reflection of the electromagnetic wave at the interface, the so-called evanescent field, decreases exponentially with the increasing distance from the interface. The distance wherein the field amplitude falls to 1/e (1/e = 0.3679...) of its value at the surface is defined as **penetration depth d_P**. Due to the decrease of the strength of the evanescent field regions just beneath the interface contribute more strongly to the spectrum than deeper ones, i.e.

weighting takes place. Since the evanescent field does not vanish at d_P, regions at distances greater than the penetration depth will also contribute to the spectrum, but with very small weight. D_P is given by (λ_0 - wavelength in vacuum)

$$d_P = \frac{(\lambda_0/n_1)}{2\pi(\sin^2\theta_i-(n_2/n_1)^2)^{1/2}} \quad \text{Eq. 8-14}$$

Eq. 8-14 leads to d_P values of about 1µm for the commonly applied experimental conditions as shown in Tab. 8-1:

Tab. 8-1: Examples of penetration depths (µm) for $\theta_i = 45°$

ATR-element	n_1 (λ = 10 µm)	Wavenumber v								
		3000 cm^{-1} (3,33 µm)			1000 cm^{-1} (10 µm)			500 cm^{-1} (20 µm)		
		n=1,4	1,5	1,55	1,4	1,5	1,55	1,4	1,5	1,55
ZNS	2,20	0,78	1,28	4,91	2,35	3,87	12,0	4,70	7,74	24,1
KRS-5	2,37	0,58	0,71	0,83	1,73	2,13	2,50	3,46	4,27	5,00
ZnSe	2,40	0,55	0,67	0,77	1,66	2,00	2,30	3,22	4,02	4,61
SI	3,40	0,27	0,28	0,29	0,81	0,85	0,87	1,63	1,70	1,73
Ge	4,00	0,22	0,22	0,22	0,65	0,66	0,67	1,30	1,33	1,35

The value 1µm is only a rough approximation of d_P, since d_P on the one hand, when approaching the critical angle, shows larger values (see table 1 for ZnS and n_2=1.55) and is on the other hand directly dependent on the wavelength. That means for inhomogeneous samples that e.g. an absorption band of a CH-stretching vibration (about 3000 cm^{-1}) caused by a layer just below the surface can be measured without overlapping with absorptions caused by deeper layers, whereas in the long wave part of the spectrum bands due to the deeper sample regions may influence or even dominate the ATR spectrum.

To understand the band intensities of an ATR spectrum, especially for the comparison with the intensities of the corresponding bands in transmission spectra, the **effective thickness** d_{eff} is an important quantity. By definition the effective thickness represents the actual thickness of the layer required to obtain the same absorption in a transmission measurement as that obtained in an ATR measurement. The effective thickness is a measure for the sensitivity of an ATR experiment. Having calculated a great effective thickness for an absorption band means a high sensitivity for this band. The dependence of the effective thickness on the experimental parameters is different for thick and thin layers and for perpendicular and parallel polarization. For weak absorbers these correlations are relatively simple [30,33].

For thick layers ($d > \lambda_1 = (\lambda_0/n_1)$) d_{eff} is directly proportional to the wavelength which means, that bands with identical absorption coefficients seem to be stronger for longer wavelengths and weaker for shorter wavelengths. In addition, absorption bands are widened on their longwave side when compared with transmission spectra. This marked diversion can be compensated when applying the software of modern spectrometers (ATR correction). The effective thickness increases with decreasing angle of incidence and decreasing ratio n_2/n_1 (this can be influenced by the choice of the ATR crystal), that means, the spectral contrast is more pronounced. Here it has to be taken into account that for all ATR experiments the condition $n_1 \sin\theta_i > n_2$ has to be fulfilled. If the experimental parameters are approaching these limits (be it through a variation of the angle of incidence θ_i or through a change of n_1, that means change of the ATR crystal) the effective thickness increases greatly (if $\theta_i \to \theta_c$ then $d_{eff} \to \infty$). This becomes noticeable in changes of the position, the intensity and the shape of absorption bands. Additionally, the refractive index of the sample undergoes major variations in the vicinity of strong absorption bands. Here n_2 can reach values of 2 or more, e.g., for polymers whose refractive indices usually remain around 1.5. Under unfavourable circumstances this might lead to a failure of the above condition for the band concerned - in contrast to the remaining spectrum. This would result in a distortion of the ATR spectrum limited to a narrow region of wavelength. Despite stronger band intensities measurements should therefore not be performed too close to the critical angle.

Measurements with minor penetration depths which can be realized by using large angles of incidence and/or large n_1 (see table 1), result in small effective thicknesses and, therefore, only weak bands. In this case, an increase of the intensities of the absorption bands in the ATR spectrum can be achieved by using an experimental setup with multiple reflections of the IR radiation at the sample.

For thin samples ($d \ll \lambda_1 = (\lambda_0/n_1)$) d_{eff} is not dependent on the wavelength, since the field strength of the evanescent field is almost constant over these small distances. Here the effective thickness of the layer is directly proportional to the thickness of the sample. Then the usable range of the angle of incidence is considerably influenced by the refractive index of the medium next to the sample (often this is air). This leads to the situation that the spectra are not distorted when the angle of incidence approaches the limiting value θ_c, determined by the ATR crystal and the sample, and even smaller values may be selected for the angle of incidence without impinging on the condition for total reflection.

Light that is polarized parallel to the plane of incidence results in larger effective thicknesses - and therefore spectra of better contrast - than perpendicular polarized light. For an angle of incidence of 45°, e.g., the relation $d_{eff,II} = 2 \cdot d_{eff,\perp}$ is valid, so that the band intensities in a non-oriented sample will differ by the factor of 2. The different interaction with the sample for parallel or perpendicular polarized light is the reason that for anisotropic objects the ATR spectrum, even when using non-polarized light, is dependent on the orientation of the sample. Detailed investigations of the orientation of anisotropic samples can be performed using ATR spectra which were recorded with polarized light [33,34,35]. In contrast to transmission measurements at perpendicular incidence, the ATR method (similar to the external reflection spectroscopy) can make orientations accessible that are perpendicular to the surface of the studied sample, since parallel polarized light comprises a component in this direction, too.

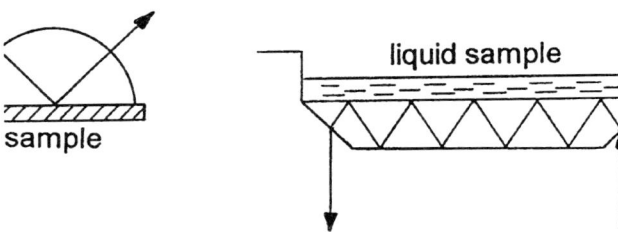

Fig. 8-13: Examples of ATR elements; a - semi-cylinder for one reflection, b - sample holder of an overhead-ATR-unit for liquid samples, multiple reflection

Today, the spectroscopist can choose from a large selection of ATR units with ATR elements made of different materials and in different shapes, often interchangeable, adaptable to the specific purposes (Fig. 8-13) [30,33]. Important criteria for the choice of material are an adequately high refractive index, high transparency in a rather large range of the IR region, insensitivity as far as chemical reactions with the sample are concerned and when being cleaned, but also mechanical properties such as hardness or brittleness. Among the preferred materials are germanium, silicon, zinc selenide, KRS 5 (see Tab. 8-2).

Most of the ATR units are supplements which can be placed in the sample compartment; there are, however, also measuring probes coupled by suitable light transfer systems to the spectrometer for external ATR measurings, e.g. in reaction containers. Under all conditions, a close contact between the ATR element and the sample is indispensible.

Solid samples are pressed against the crystal for this purpose, (care needs to be taken with hard samples: there is a danger of scratching soft crystals or breaking brittle ones).

Tab. 8-2: Properties of some ATR elements

ATR-element	θ_c (for $n_2 = 1.5$)	Transmiss.-range cm^{-1}	Remarks
ZnS	43,0	22000-750	unsoluble in water
KRS-5	39,3	16000-300	poisoneous
ZnSe	38,7	20000-500	rel. unsoluble
Si	26,2	8300-33	very hard
Ge	22,0	5000-600	very brittle

When investigating liquid samples, the ATR element is rinsed by the liquid (partly immersed even), or the liquid is put on a crystal that is horizontally fastened. When in cases with solutions or dispersions the liquid medium interferes in the spectrum one can try to eliminate this influence by subtracting

its spectrum or by use of the liquid as background. Often simple evaporation of the liquid component followed by appropriate drying of the residue or the resulting film proves helpful. There are a large number of ways how to apply thin layers on ATR crystals; just mentioning dipping, casting and spin-coating can play a large part; surface tension may be important then [19,35,36,37].

ATR units with multiple reflection are preferred and used with success for studies aimed at qualitative results because of their larger band intensities. ATR units with only one reflection are preferable for exact quantitative measurements, possibly with the determination of the optical constants, since the divergence of the light beam causes less deteriorating effects in this case. The required amount of substance is larger with multiple reflection units due to the generally needed larger surface of the sample. Lately, for IR microscopes ATR objectives were developed which allow an ATR investigation of smallest sample matter, e.g. of microscopic enclosures in a matrix. The minimal lateral object dimensions amount to some ten micrometers.

The ATR method offers the opportunity of determining depth profiles in objects that are inhomogeneous as far as their chemical composition or their physical structure is concerned [38-41]. The necessary variation of the penetration depth can be achieved by a change of the ATR crystal (variation of n_1) or by changing the angle of incidence θ_i. Larger n_1 or θ_i will result in less penetration. For these purposes crystals of the different materials with differently bevelled edges are available. Here a choice of ATR units with adjustable angle of incidence is especially recommended since the need for repositioning the sample and the accompanying danger of adding mistakes is eliminated. You can also vary the depth of penetration by placing a thin layer of a suitable material, transparent at least in the relevant range of the spectrum, between the ATR crystal and the sample [42].

UNIVERSITY OF MAINE
Fogler Library

DATE DUE

DEC 2 7 1999

SEP 0 1 2000

- Return or renew materials on or before date stamped above.

- Items may be renewed by telephone (581-1666).

- Overdue fines accumulate at $.50 per day.

- Replacement cost charged for items not returned.

- Do not rely on overdue notices to avoid charges. You are responsible for overdue material whether or not you receive an overdue notice.

- Items subject to Recall after two weeks.

The following figures present some examples of ATR spectra of polymer latices and the information contained therein. Fig. 8-14 compares the ATR spectrum of a core-shell-latex (c) with the ATR spectra of two latices having the same chemical composition as the polymer material in the shell (a) or the core (b), respectively.

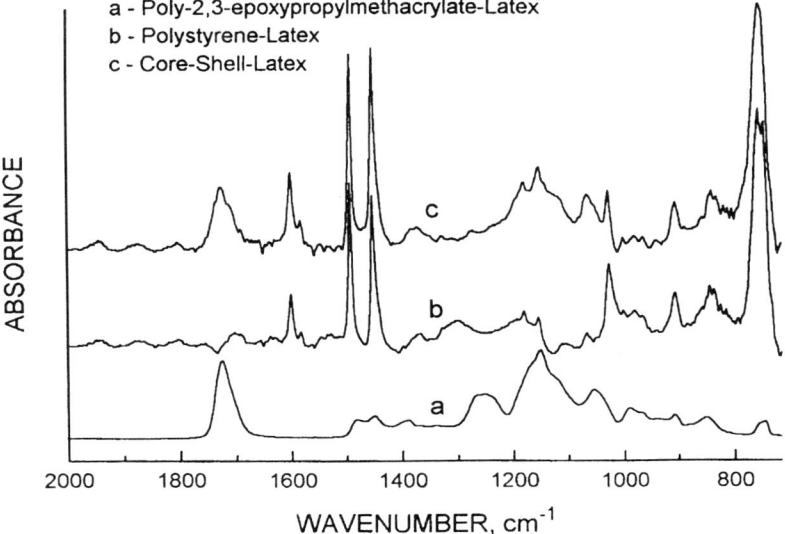

Fig. 8-14: Qualitative comparison of the ATR spectra of different polymer latices dried on a Ge-ATR-crystal;
a: poly-2,3-epoxypropylmethacrylate latex, b: polystyrene latex, c: core-shell-latex (identical with Fig. 8-6). Different ordinate units of the three spectra.

Obviously the spectrum of the core-shell-latex is composed of the absorptions due to the core on the one hand and the shell on the other hand. As a consequence of different weighting it is, however, not the simple sum of these two components. Furthermore, after drying the three latices differ considerably concerning the contact with the ATR element so that the absolute intensities of the bands in the three spectra must not be compared directly.

Fig. 8-15 compares the transmission spectrum of a core-shell-latex (a) consisting of a polystyrene core and an acrylate shell with the ATR spectra of the same sample before (b) and after (c) ATR correction.

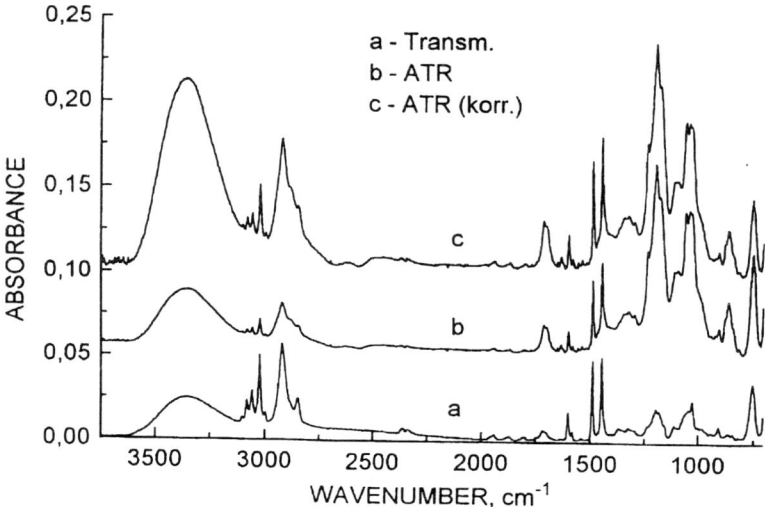

Fig. 8-15: ATR correction and comparison with the transmission spectrum (a) of the latex; b,c: ATR spectra of the latex dried on a Ge-ATR-element before (b) or after (c) ATR correction

The general intensity decrease of the bands towards shorter wavelengths can be clearly seen in the uncorrected spectrum. This effect has been cancelled in (c) so that the changed intensity relations in the ATR spectrum compared with those in the transmission spectrum become more pronounced. An example for this is given by the intensity relation of the carbonyl band of the acrylate in the shell (1717 cm^{-1}) to the polystyrene absorption of the core at 1600 cm^{-1}. This is, in accordance with the mass portions, considerably smaller than 1 in the transmission spectrum, whereas the acrylate band is <u>stronger</u> than the polystyrene band in the ATR spectrum. The reason for this the smaller distance of the acrylate to the ATR element resulting in stronger interaction with the evanescent field. The increased sensitivity of the ATR method towards surface effects is also demonstrated by the intense OH band at about 3350 cm^{-1} in spectrum (c). This band is caused by OH groups bound to acrylate molecules at or near the particle surface (see Fig. 8-7). The absorption due to these groups is much weaker in the transmission spectrum.

The possibility of in-situ-studies using a liquid-sample holder is shown in Fig. 8-16. It reproduces ATR spectra recorded during gradual drying of a latex. First spectra are strongly dominated by the absorptions of the dispersion agent water (broad bands at 3400 cm^{-1}, 1640 cm^{-1}, 700 cm^{-1}), but with decreasing water content the bands of the polymer material become more and more clearer so that they can be studied in detail.

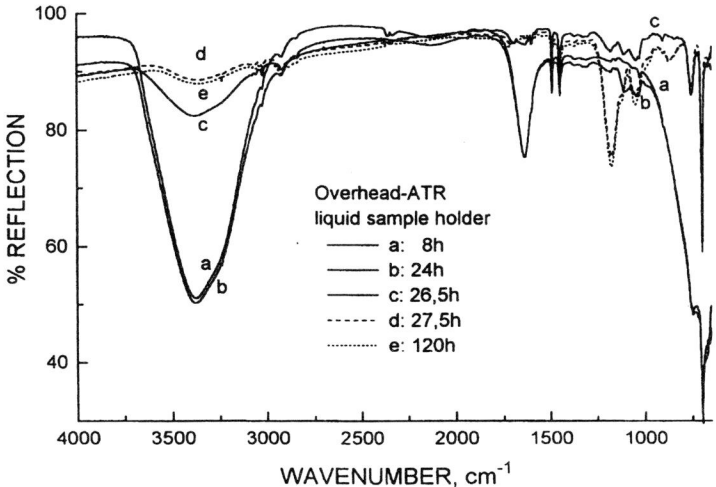

Fig. 8-16: ATR spectra of a polymer latex with polystyrene and acrylate components recorded during drying

8.4.3 Diffuse Reflection

A further infrared spectroscopic research technique is the diffuse reflection. It utilizes the portion of radiation scattered by the sample, i.e. it measures radiation which has a disturbing effect in case of other spectroscopic techniques, especially the transmission spectroscopy.

In case of interaction of infrared radiation with the scattering sample (powder, surface of a solid, and others) the radiation hits the surface of the particles under varying angles of incidence. Hence some rays are specularly reflected, others are refracted and propagate within the particles. The latter may leave the particles lest they be totally reflected within the particles again (Fig. 8-17).

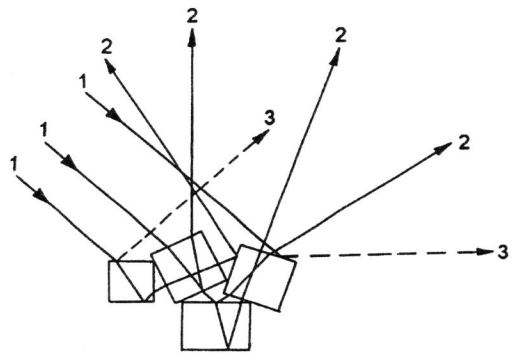

Fig. 8-17: Principle of diffuse reflection. 1: incident beam, 2: multiple reflected rays (interaction with volume as well as with surfaces of particles), 3: specularly reflected rays (interaction with particle surfaces)

Several of those processes may occur subsequently before this portion of the radiation leaves the sample under various angles. The intensity of the radiation penetrating into the particles of the sample is weakened by absorption processes contrary to the specularly reflected radiation. Both portions of the radiation are distributed in all directions of the space above the sample. Its intensity is recorded in dependence on the wave number.

The resulting spectra are only to some extent similar to transmission spectra of equal sample material. This is partly due to the different pathlengths which the rays have to pass within the sample, or to the mixing of the different kinds of the reflected radiation outlined above. As a consequence of these influences the band intensities do not differ greatly ("levelled" spectrum) or the spectra are perturbed by dispersion effects when compared to transmission spectra [43]. Therefore the spectra are usually transformed into a shape allowing its interpretation in analogy to transmission spectroscopy. Then the ordinate readings in absorbance units are predominantly proportional to the extinction coefficient and the concentration.

A model first described by Kubelka and Munk [44] and comprehensively elaborated upon by Kortüm [45] has realized such transformation using the moduli K and S (**K: absorption module, S: scattering module**) which are dependent on wavenumber.

The Kubelka-Munk model presupposes isotrope scattering. The sample is described as semi-infinite continuum, which is to say that it spreads infinitely in the plane of its surface and in one direction perpendicular to it. When imagining such sample dissected into separate layers parallel to its surface the flux of radiation through one layer can be calculated. In this procedure portions of the radiation are included which reach the layer considered from the layers above as well as from the layers below showing isotropic angle distribution. The summation of all layers of the sample results in the finding of the reflection degree R_∞ of the infinitely thick layer which is defined as the proportion of the outcoming and incoming flux of radiation. It is related to the absorption and the scattering moduli respectively as shown in Eq. 8-15.

$$F(R_\infty) \equiv (1-R_\infty)^2/2R_\infty = K/S \qquad \text{Eq. 8-15}$$

with $F(R_\infty)$ being the Kubelka-Munk function. For the absorption module K is valid [43]:

$$K = 2 \cdot \ln 10 \cdot \varepsilon \cdot c \ldots \qquad \text{Eq. 8-16}$$

In many experiments the scattering module S was constant. In those cases the Kubelka-Munk function is directly proportional to the molar decadic extinction coefficient and to the concentration. Diffuse reflection spectra shown in Kubelka-Munk units, $F(R_\infty)=f(\nu)$, can be evaluated qualitatively (e.g. library search) and quantitatively similar to transmission spectra.

However, particular sources of error are to be considered. They are caused by the fact that some presuppositions of the Kubelka-Munk model are not met in practice [43].

Similar to the ATR technique, e.g., the experimental decrease of the flux penetrating into the sample results in major contributions to the intensity measured coming from the uppermost layers of the sample. But these layers are irradiated under practically the same angle so that the isotropic angle distribution as used in the Kubelka-Munk model is not yet achieved. Particularly for strongly absorbing samples this model is not free of errors since the infrared radiation reaches only the top layers of such samples.

Another source of errors is caused by that part of the measured intensity which has been specularly reflected. It results in dispersion-like effects (see section 3.1) in the spectra. This effect is found especially with samples of defined surface, but also with powdery samples where specular reflection occurs at the surfaces of the particles. With powdery samples the errors are minimized by dilution of the sample in a minimally absorbing, powdery matrix (e.g. potassium bromide). Samples with defined surfaces, however, require the exclusion of the range of reflection angles comprising specularly reflected light to eliminate this disturbing influence [46]. In case of dilution of the sample in an absorption-free matrix the matrix material itself is used as reference in determining the reflection degree of the sample. However, absorption-free, scattering reference samples cannot be exactly achieved in practice. Particularly contaminations cause small absorptions, yet they have a considerable impact on the determination of the reflection degree. Potassium bromide or sodium chloride are widely used as reference materials and enable in many cases a sufficiently dependable qualitative and quantitative analysis in the mid-infrared [47].

In principle, the diffuse reflection spectroscopy produces similar results as the transmission spectroscopy. The transformation of measured spectra into Kubelka-Munk units can be achieved by the software which is provided together with modern spectrometers. Nevertheless, there are limitations in the use of the diffuse reflection spectroscopy. Difficulties may occur in band assignment and in quantitative analysis of the spectra. Mastering these difficulties often requires time-consuming systematic studies. In spite of that the diffuse reflection spectroscopy has spread widely. It is particularly used in the near infrared spectral region and is gaining application in the mid-infrared especially for research in powder samples. Due to the non-problematic or even unnecessary sample preparation the diffuse reflection spectroscopy is particularly interesting for industrial use. It is used in chemistry, material testing, and quality control. In the pharmaceutical industry it is applied in characterizing the components of pills.

Diffuse reflection spectroscopy is applicable to the investigation of surfaces, since the detected radiation comprises predominantly information from the uppermost layers of the sample, or it has interacted repeatedly with surfaces within the sample in case of powders [48]. Samples with large specific surfaces (powders, zeolithes, catalysts,latices etc.) can be analysed successfully. Fig. 8-18 shows the diffuse reflection spectrum of a

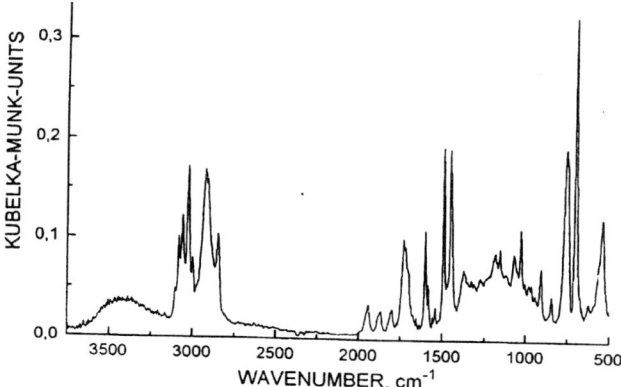

Fig. 8-18: Diffuse reflection spectrum of a dried polymer latex (same latex as in Fig. 8-6)

dried latex sample prepared by dilution in a KBr matrix (The diffuse reflection spectra shown were recorded in the Bundesanstalt für Materialforschung und -prüfung (Dr. Reklat)).

Characteristic bands indicate the core-shell-structure of the material (core: polystyrene; bands at 1600 cm^{-1}, 1585 cm^{-1}, 1500 cm^{-1}, 1460 cm^{-1}, 760 cm^{-1}, 700 cm^{-1}; shell: epoxyacrylate, 1720 cm^{-1}). The absorption bands of the structural elements at the surface of the latex particles can be clearly seen (e.g. stretching vibration of OH groups at about 3430 cm^{-1}). The spectrum illustrates the effectiveness of the diffuse reflection spectroscopy in studying particle surfaces. Subjects of investigation are chemical composition, structure and orientation of surface coatings as well as reactions on surfaces which are caused by special treatments, influences of surrounding media or fatigue processes.

8.5 References

[1] E. G. Brame, J. G. Grasselli: "Infrared and Raman Spectroscopy", Practical Spectroscopy Series, Vol.1, Part A,1976; Part B,1977; Part C, 1977; Marcel Dekker, Inc., New York, Basel

[2] Harrick Scientific Corporation (Ed.):" Optical Spectroscopy: Sampling Techniques Manual", Ossining, NY, 1987

[3] R.G. Messerschmidt, M.A. Harthcock: "Infrared Microspectroscopy - Theory and Applications", Practical Spectroscopy Series, Vol.6, Marcel Dekker, Inc., Basel, 1988

[4] N.J. Everall, J.M. Chalmers, A. Local, S. Allen: "Measurement of surface orientation in uniaxial poly(ethylene terephthalate) films using polarised specular reflectance Fourier transform infrared microscopy", Vibr. Spectroscopy, 10 (1996), 253-259

[5] H. Günzler, H.Böck: "IR-Spektroskopie - eine Einführung", Verlag Chemie GmbH, Weinheim, 1983

[6] J. Dechant: Ultrarotspektroskopische Untersuchungen an Polymeren", Akademie-Verlag, Berlin, 1972

[7] H.W. Siesler, K. Holland-Moritz: "Infrared and Raman Spectroscopy of Polymers", Practical Spectroscopy Series, Vol.4, Marcel Dekker, Inc., New York, Basel, 1980

[8] D. Lin-Vien, N.B. Colthup, W.G.Fateley, J.G.Grasselli: "The Handbook of Infrared and Raman Characteristic Frequencies of Organic Molecules", Academic Press, Inc. San Diego, CA; London, 1991

[9] G. Socrates: "Infrared Characteristic Group Frequencies - Tables and Charts", John Wiley & Sons Ltd., Chichester, New York, 1994

[10] D.O. Hummel, F. Scholl: "Atlas of Polymer and Plastics Analysis", Vol.1, 1978; Vol.2, Part a, 1984, Part b, 1988; Vol.3, 1981, Carl Hauser Verlag, München; Verlag Chemie, Weinheim

[11] K.G.R. Pachler, F. Matlock, H.-U.G. Gremlich: "Merck FT-IR Atlas - a Collection of FT-IR Spectra", VCH Verlagsgesellschaft mbH, Weinheim, 1988

[12] B. Schrader: "Raman/Infrared Atlas of Organic Compunds", VCH Verlagsgesellschaft mbH, Weinheim, 1989

[13] S. Kawata, K.Takeuchi: "Infrared thin-film analysis with two transmission spectra measured at different incident angles", J.Opt.Soc.Am.A, Vol.8 (1991), 1055-1061

[14] Y. Nishikawa, K. Fujiwara, T. Shima: "A Study on the Qualitative and Quantitative Analysis of Nanogram Samples by Transmission Infrared Spectroscopy with the Use of Silver Island Films", Appl. Spectroscopy, Vol.45 (1991), 747-751

[15] J.C. Henniker:" Infrared Spectrometry of Industrial Polymers", Academic Press, London, New York, 1967

[16] J.P. Hawranek, R.N. Jones: "The control of errors in i.r. spectrophotometry - V. Assessment of errors in the evaluation of optical constants by transmission measurements on thin films", Spectrochimica Acta, Vol. 32A (1976), 99-110

[17] J.E. Bertie, C.D. Keefe, R.N. Jones: "Infrared intensities of liquids VIII. Accurate baseline correction of transmission spectra of liquids for computation of absolute intensities, and the 1036cm^{-1} band of benzene as a potential intensity standard", Can. J. Chem., Vol.69 (1991), 1609

[18] K. Yamamoto, H. Ishida: "Optical Theory Applied to Infrared Spectroscopy", Vibr. Spectroscopy, Vol.8 (1994), 1-36

[19] E. Kientz, Y. Holl: "Distribution of surfactants in latex films", Colloids & Surfaces A-Physicochemical & Engineering Aspects, Vol.78 (1993), 255-270

[20] M. Born, E.Wolf: "Principles of optics", Pergamon Press, Oxford, 1980

[21] E.H. Korte: "Infrared Specular Reflectance of Weakly Absorbing Samples", Vibr. Spectroscopy, Vol.1 (1990), 179-185

[22] A.N. Parikh, D.L. Allara: "Quantitative Determination of Molecular Structure in Multilayered Thin Films of Biaxial and Lower Symmetry from Photon Spectroscopies.I. Reflection Infrared Vibrational Spectroscopy", J.Chem.Phys., Vol.96 (1992), 927-945

[23] M. Handke, M. Milosevich, N.J. Harrick: "External Reflection Fourier Transform Infrared Spectroscopy: Theory and Experimental Problems", Vibr. Spectroscopy, Vol.1 (1991), 251-262

[24] Y. Ishino, H. Ishida: "FT-IR External Reflection Spectroscopy at Brewster's Angle", Appl. Spectroscopy, Vol.46 (1992), 504-509

[25] J.A. Mielczarski, E. Mielczarski: "Determination of Molecular Orientation and Thickness of Self-Assembled monolayers of Oleate on Apatite by FTIR Reflection Spectroscopy", J.Phys. Chem., Vol.99 (1995), 3206-3217

[26] A. Gericke, A.V. Michailov, H. Hühnerfuss: "Polarized External Infrared Reflection-Absorption Spectrometry at the Air/Water Interface: Comparison of Experimental and Theoretical Results for Different Angles of Incidence", Vibr. Spectroscopy, Vol.4 (1993), 335-348

[27] N.J. Harrick: "External Reflection Spectroscopy: Selection of Optimum Parameters for Thin Films on Smooth Metal Surfaces", American Laboratory, Nov. 1986, 78-81

[28] F. Peñacorada, J. Reiche, R. Dietel, T. Zetzsche, B. Stiller, H. Knobloch, L. Brehmer: "Monolayers and Multilayers of Uranyl Arachidate. 2.Influence of the Subphase pH on the Structure and Stability of Langmuir-Blodgett Films", Langmuir, 1996, in press

[29] T. Kawai, J. Umemura, T. Takenake: "Molecular Orientation in LB-Films of Azobenzene-Containing Long-Chain Fatty Acids and Their Barium Salts Studied by FT-IR Transmission and Reflection Spectroscopy", Langmuir, Vol.6 (1990), 672-676

[30] F.M. Mirabella: "Internal Reflection Spectroscopy - Theory and Applications", Practical Spectroscopy Series, Vol.15, Marcel Dekker, Inc., New York, Basel, 1993

[31] B.J. Niu, M.W. Urban: "Surface and interfacial FTIR spectroscopic studies of latexes.9. The effect of homopolymer and copolymer structures on surfactant mobility in STY/BA latices", J.Appl.Polymer Sci., Vol.56 (1995), 377-385

[32] E.K. Kemsley, G.P. Appleton, R.H.Wilson: "Quantitative analysis of emulsions using attenuated total reflectance (ATR)", Spectrochimica Acta Part A-Molecular Spectroscopy, Vol.50 (1994), 1235-1242

[33] N.J. Harrick: "Internal Reflection Spectroscopy", John Wiley & Sons, Inc., New York, 1967

[34] F.M. Mirabella jr.,N.J.Harrick: "Internal Reflection Spectroscopy: Rewiew and Supplement", Marcel Dekker, Inc., New York, Basel, 1985

[35] J.P. Kunkel, M.W.Urban: "Surface and interfacial FT-IR spectroscopic studies of latexes.8. The effect of particle and copolymer composition on surfactant exudation in styrene-n-butyl acrylate copolymer latex films", J. Appl.Polymer Sci., Vol.50 (1993), 1217-1223

[36] T.A. Thorstenson, M.W. Urban, "Surface and interfacial FTIR spectroscopic studies of latexes.5. The effects of copolymer composition on surfactant exudation", J.Appl.Polymer Sci., Vol.47 (1993), 1387-1393

[37] C.M. Balik, J.R. Xu: "Simultaneous measurement of water diffusion, swelling, and calcium carbonate removal in a latex paint using FTIR-ATR", J.Appl.Polymer Sci., Vol. 52 (1994), 975-983

[38] J.B. Huang, M.W. Urban: "Novel approach to quantitative depth profiling of surfaces using ATR/FT-IR measurements", Appl. Spectroscopy, Vol.47 (1993), 973-981

[39] R.A. Shick, J.L. Koenig, H. Ishida: "Theoretical Development for Depth Profiling of Stratified Layers Using Variable-Angle ATR", Appl. Spectroscopy, Vol.47 (1993), 1237-1244

[40] J.G. Van Alsten: "Diffusion Measurements in Polymers Using IR Attenuated Total Reflectance Spectroscopy", TRIP, Vol.3 (1995), 272-276

[41] L.J. Fina, G. Chen: "Quantitative depth profiling with Fourier transform infrared spectroscopy", Vibr. Spectroscopy, Vol.1 (1991), 353-361

[42] E. Marand,L.M. Smartt: "ATR Spectroscopic Study of PMMA/PDMS Graft Copolymers Using a Barrier Layer Method", Appl. Spectroscopy, Vol.49 (1995), 513-519

[43] E.H. Korte: "Infrarot-Spektroskopie diffus reflektierender Proben" in: H.Günzler et al. (Ed.), Analytiker-Taschenbuch, Bd.9, 91-123

[44] P. Kubelka, F. Munk: "Ein Beitrag zur Optik der Farbanstriche", Z. techn. Physik, Vol.12 (1931), 593-601

[45] G. Kortüm: "Reflexionsspektroskopie", Springer-Verlag, Berlin, 1969

[46] E.H. Korte, A.Otto: "Infrared Diffuse Reflectance Accessory for Local Analysis on Bulky Samples", Appl. Spectroscopy, Vol. 42 (1988), 38-43

[47] D. Reineke, A. Jansen, F. Fister, U. Schernau: "Quantitative Determination of Organic Compounds by Diffuse Reflectance Fourier Transform Infrared Spectrometry", Anal. Chemistry, Vol.60 (1988), 1221-1224

[48] J.A.J. Jansen, W.E. Haas: "Applications of Diffuse Reflectance Optics for the Characterization of Polymer Surfaces by Fourier Transform Infra-Red Spectroscopy", Polymer Communications, Vol.29 (1988), 77-80

Addresses of the authors:

Dr. A. Büchtemann
Fraunhofer Institute of Applied
Polymer Research
Kantstr. 55
14513 Teltow-Seehof

Dr. R. Dietel
University Potsdam
Research Group "Thin Organic Films"
Kantstr. 55
14513 Teltow-Seehof

9 Circular Dichroism (CD) for the Analysis of Dissolved and Adsorbed Proteins

Dr. H. Hermel, Max-Planck-Institut, Berlin-Adlershof

9.1 General Remarks [1-3]

Circular dichroism results from the interaction of a given optically active medium with linearly polarized light. As it is well known, linearly polarized light is a superposition of two left and right circularly polarized components of equal amplitude (Fig. 9-1a). What is observed due to the interaction?

1. In the spectral range where the medium does not absorb an optical rotation α is observed, because the refractive indices n_l and n_r are different, resulting in different light velocities for both components (Fig. 9-1b). The optical rotatory dispersion (ORD) curve $\alpha = f'_{(\lambda)}$ increases with decreasing wavelength (Fig. 9-2, normalous dispersion), as it is expressed by the Drude-equation

$$\alpha = A / (\lambda^2 - \lambda_i^2) \qquad A, \lambda_i \text{ are constants} \qquad \text{Eq. 9-1}$$

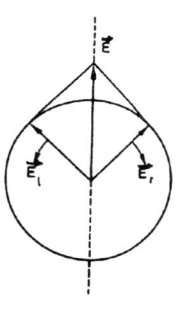
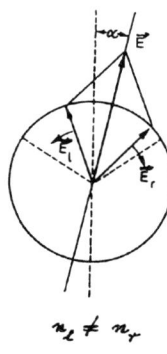
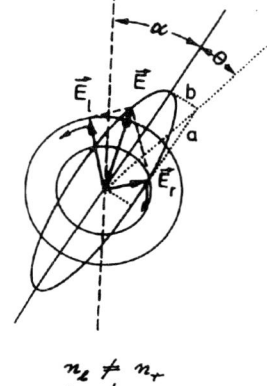

Fig. 9-1: Projection of the electric field vectors showing the composition of linearly polarized light.
a) Two equal components of left and right circularly polarized light.
b) Electric field vectors after irradation of the optically active medium without absorption.
c) Electric field vectors after irradation of the optically active medium with light in the absorption range.

2. If the medium has an absorption band, the left and the right circularly polarized components have not only different velocity but also different absorption, e.g. the left more than the right. Elliptically polarized light results (Fig. 9-1c). The ORD curve exhibits anomalous dispersion (Fig. 9-2). The circularly dichroic absorption (CD) coefficient $\Delta\varepsilon$ is defined as the difference $\varepsilon_l - \varepsilon_r$. The maximum of the CD curve corresponds to the reversal point of the anomalous ORD curve. Both effects, CD and anomalous ORD are called Cotton-effects after their discoverer.

A positive Cotton-effect exhibits positive rotation on the long wavelength side of the crossover point and a negative effect on the short wavelength side. The Cotton-effect, including both anomalous ORD and CD, is based on

electron transitions	$n \to \pi^*$	UV/Vis
	$\pi \to \pi^*$	
molecule vibrations		IR

A quantitative measure of the CD-absorption band is the integral circulardichroitic absorption. This is proportional to the rotational strength R of the electron transition. As Eq. 9-2 indicates, a nonzero rotational strength requires therefore, that the electric and magnetic transition moments are not perpendicular to each other.

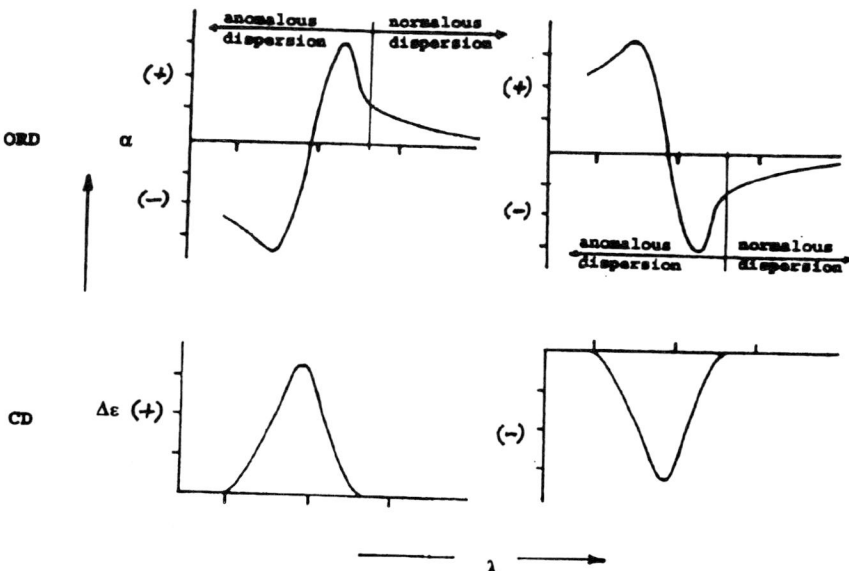

Fig. 9-2: Comparison of ORD- and CD-spectra

$$\int_{\lambda_2}^{\lambda_1} \frac{\Delta\varepsilon(\lambda)}{\lambda} d\lambda = R = \mu_e \cdot \mu_m \cdot \cos(\vec{\mu_e}, \vec{\mu_m})$$

Eq. 9-2

If R can be calculated ab initio it is possible to determine the sign of the Cotton-effect. For a positive CD $\Sigma\mu_e,\mu_m$ have to be parallel, whereas in the case of antiparallel $\Sigma\mu_e,\mu_m$ the CD is negativ.

9.2 Origin of Absorption and Dispersion

Optically active molecules are chiral. If they have a chromophore then a Cotton-effect will be exhibited and CD can be observed. A list of principal chromophores is shown in Fig. 9-3. Two different cases may occur: The chromophore itself is dissymmetric,

Chromophore	Transition	λ max nm
>C=C-C=C<	π → π*	240
>C=O	n → π*	260-270
H>C=O	n → π*	300
>C=C-C=O	π → π* n → π*	240 330
>C=N-	n → π*	250
>C=S	π → π* n → π*	235-240 500
-CO-O-	n → π*	210-220
-CO-N- polypeptide	π → π* n → π*	
CO-S-	π → π* n → π* (?)	235 270
-O-CS-S	π → π* n → π*	280 355
-N-CS-S	π → π* n → π*	280 330
-C-S-C-	n → π* (?)	260
-SO-	–	210
-NO₂	n → π*	280
-NO	n → π*	680
-ONO	?	350-400

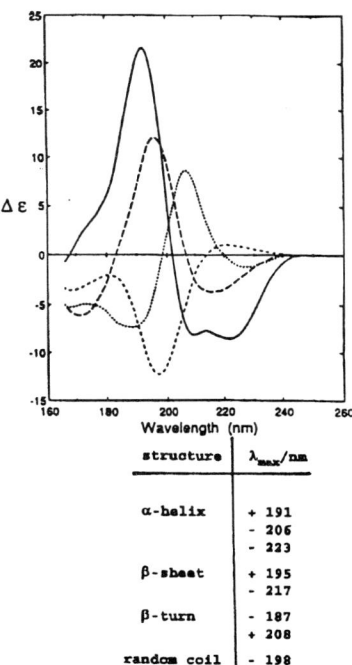

structure	λ_{max}/nm
α-helix	+ 191 - 206 - 223
β-sheet	+ 195 - 217
β-turn	- 187 + 208
random coil	- 198 + 217

Fig. 9-3: Principal chromophores and the CD-spectra for various secondary structures of polypeptides.
α-helix (——), antiparallel β-sheet (——), β-turn (····) and random coil (---).
(Polypeptide-CD spectra from: S. Brahms, J. Brahms, J. Mol. Biol. 1980, 138, 149).

then the resulting CD is very strong, or the chromophore is symmetric, but contained in a dissymmetric molecule, then the optical activity is induced in the chromophore by its environment, which results in a much weaker CD.

The chromophore of the polypeptides and proteins is the -CO-N- backbone in its different secondary structure modes [4] (Fig. 9-3).

9.3 Poly-L-lysine CD, a Standard for the Secondary Structure

Poly-l-lysine occurs in water solution in 100% as α-helix as well as β-sheet or random coil in dependence of the conditions in the solution (Fig. 9-4a). Therefore it is a good CD-standard for the polypeptide and protein secondary structure (Fig. 9-4b). For example, Fig. 9-4c and Fig. 9-4d show the CD spectra of mixtures of different secondary structures. This is the classical work of Greenfield and Fasman [5,6]. Nowadays the CD-spectra analysis is performed using data banks and computer programs, as shown in Fig. 9-5, giving the CD-spectra analysis of a lipoprotein [7] as an example.

9.4 Problems of CD-investigations of Adsorbed Proteins

The sensitivity of modern CD spectrometers is satisfactory enough to measure protein solutions of 0.02 mg/ml. But what are the conditions in the case of polypeptides and proteins embedded in a membrane or spread as a monolayer at an interface? In these cases the protein concentration is too small for direct CD-measurements in the usual way. Several groups have tried to overcome the problem of sensitivity by extraction and concentration processes of the monolayers, often leading to complicated sample preparation procedures. Two examples are given here:

1. The proteins are embedded in the membrane as in the investigations of Zahler et al. [8]: They studied the secondary structure of intrinsic membrane proteins and membrane associated proteins of mitochondria by CD. They have manipulated the biomaterial intensively with mechanical shearing and different extraction processes, in some cases even 8M urea was used (Fig. 9-6). Finally the protein secondary structure was detected by CD in solutions of the extracts. Changes of the secondary structure that are due to sample preparation, cannot be excluded .
2. A protein monolayer can be adsorbed at the fluid interface. Cornell [9] and Erokhin et al. [10] demonstrated this with monolayers of polyglutamate, polyalanine and cytochrome C adsorbed at the air/water interface after spreading and compressing

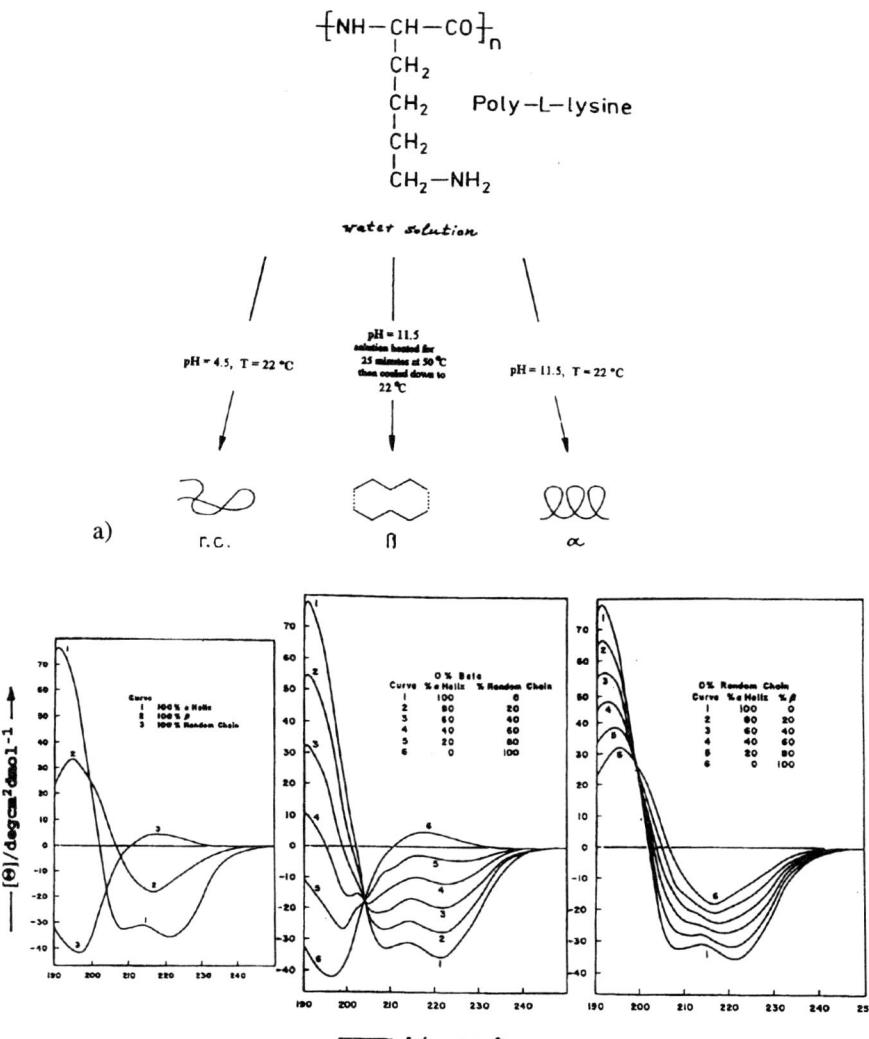

Fig. 9-4: Poly-L-lysine as a CD-standard for the different protein secondary structures.
a) The conditions of the water solution for the occurrence of different secondary structures.
b) The CD-spectra of the different secondary structures.
c),d) Calculated CD-spectra of mixtures containing varying percentages of different secondary structures.
(b),c),d) from: N. Greenfield, G.B. Fasman, Biochemistry 1969, 8, 4108).

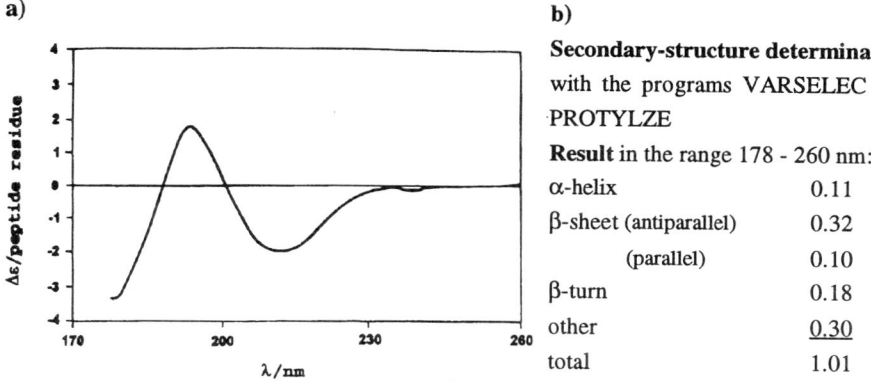

Fig. 9-5: Analysis of the secondary structure of a lipoprotein.
   a) CD-spectra in water solution at pH 7.
   b) Result of the secondary structure determination (from: L.L. France et al. Biochim. Biophys. Acta 1992, 1120, 59).

on a Langmuir-Blodgett through (Fig. 9-7). Their aim was to study the polypeptide conformation of the layer by CD. Therefore they transferred the compressed monolayer onto quartz plates with the film lift technique. Several coated plates (8 to 10) were aligned subsequently in the light beam of the CD-spectrometer with their directions of film transfer adjusted parallel in order to obtain sufficient layer thickness. All these methods, however, do not only involve complicated preparation techniques, but also have the disadvantage of possible changes of the protein structure due to the treatment. Therefore, a method for in situ measurements at the fluid interface is desirable to obtain direct information about the protein monolayers at the interface.

9.5 In situ Measurements of Adsorbed Protein Layers at the Fluid Interface

In situ measurements of adsorption layers at the interface by reflection-absorption spectroscopy under grazing beam incidence were attempted by several laboratories in the last years in the range of molecule vibration bands, which is the IR-region (IRRAS). The data and the curves in Fig. 9-8 show characteristic spectral positions for different peptide backbone conformations. The main problems for the measurements are:
- the small reflectivity of the fluid interface
- the low absorbance of the monolayer
- the very strong absorbance of water in the wavenumber region of interest (the middle IR).

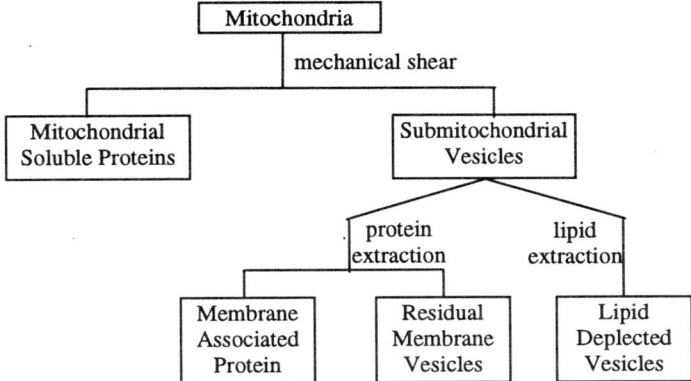

Fig. 9-6: Flow diagram for the preparation of protein samples from beef heart mitochondria for the purpose of CD-measurements (from: W.L. Zahler et al. Biochim. Biophys. Acta 1972, 255, 365).

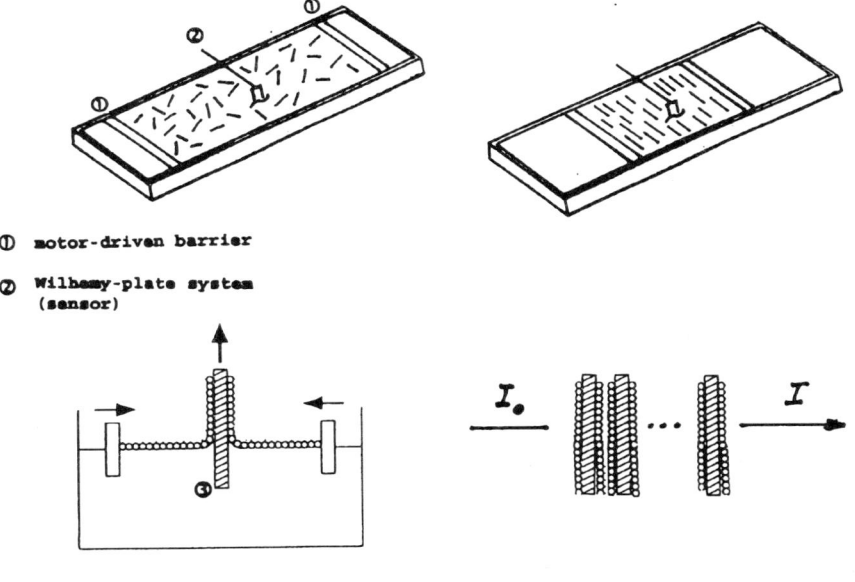

① motor-driven barrier
② Wilhelmy-plate system (sensor)
③ quartz-plate

Fig. 9-7: Scheme of a protein monolayer at the air/water interface.
    a) Protein molecules spread on the Langmuir-Blodgett through.
    b) The compressed and liquid-crystalline ordered protein monolayer.
    c) Transfer of the compressed monolayer onto quartz plates with the film-lift technique.
    d) Several coated plates in the light beam.

Conformation	Wavenumber/cm⁻¹	
	Amide I C=O stretching	Amide II C-N stretching N-H bending
α-helix	1660-1647	1550-1516
β-sheet (parallel)	1648-1632	1550-1530
(antiparallel)	1668-1632	1550-1540
random coil	1658	1535

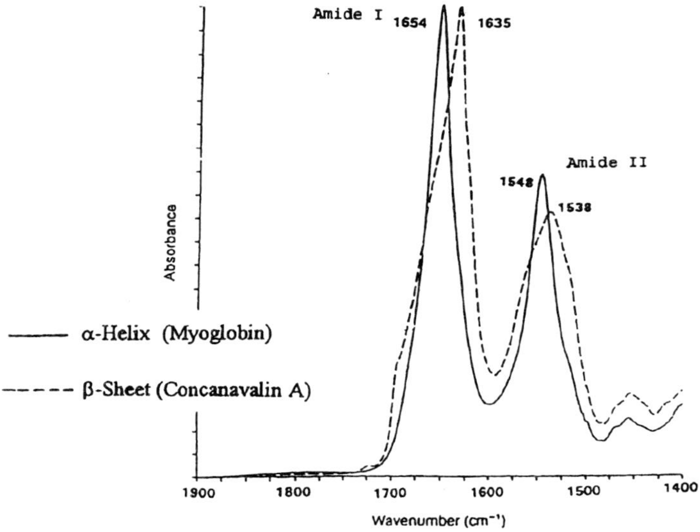

Fig. 9-8: Characteristic spectral positions in the middle IR of different polypeptide backbone conformations.

Therefore, to obtain the necessary signal/noice ratio of 10^{-6} in the case of measurements at the air/water interface, the following setup was used [11,12]: A linearly polarized IR-beam strikes the interface at grazing angle (Fig. 9-9a). The beam polarisation is modulated between the perpendicular (s) and parallel (p) state with a frequency in the kHz-range, induced by a photoelastic modulator (PEM). This is a piezo-electrically regulated octahedral ZnSe crystal with $\lambda/4$-modulation (Fig. 9-9b).

Left and right circularly polarized light results (Fig. 9-9c) with a rapidly changing resulting polarization. Therefore it is possible to measure both circular components quasi-simultaneous.

Briefly, the basic principle of this PEM-IRRAS method is to combine
- the experimental conditions for in situ reflection/absorption measurements at the air/water interface with

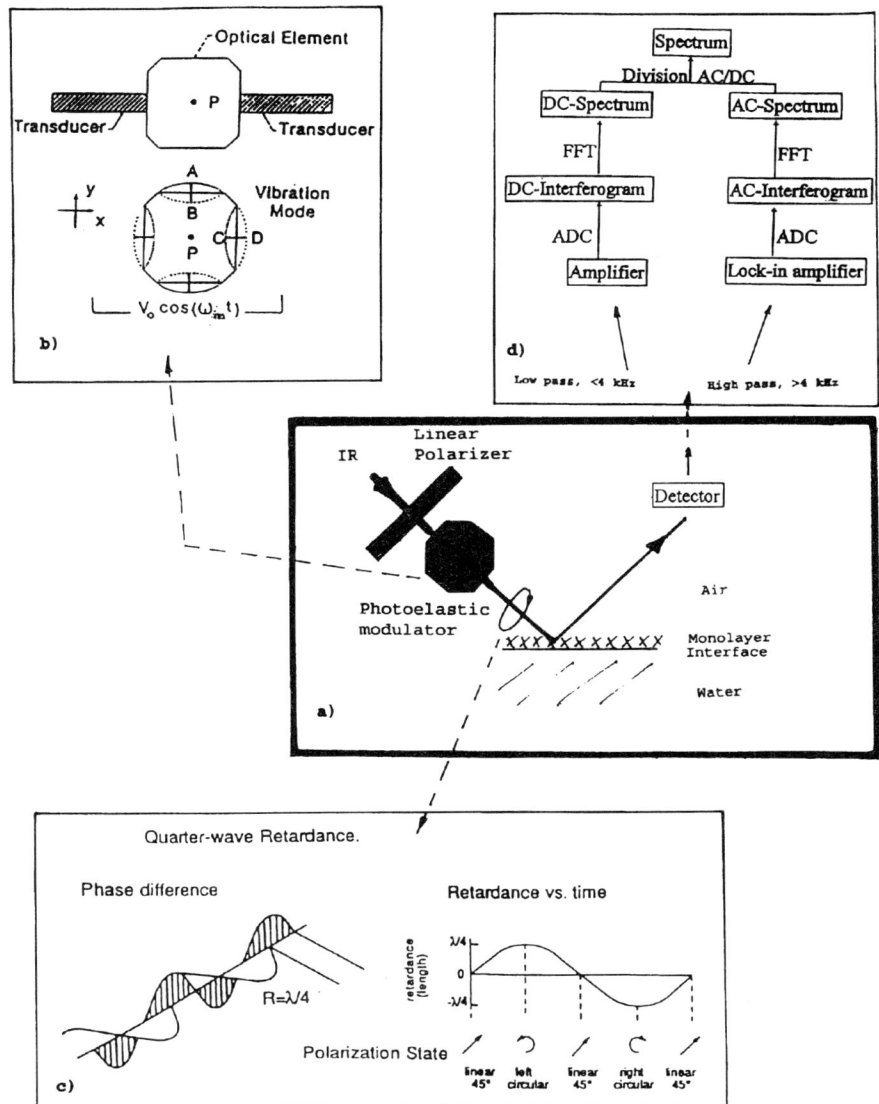

Fig. 9-9: Scheme of the polarization modulated Fourier-transform IR spectroscopy of protein monolayers at the air/water interface.
  a) Polarized reflectivity at grazing incidence.
  b) Function of the photoelastic modulator.
  c) Polarization of the incoming light beam.
  d) Block diagram of the signal processing.

- a fast modulation of polarization state of the incident electromagnetic field and
- to extract from the detected intensity, using electronic filtering and demodulation, the two signals $(R_p - R_s)$ and $(R_p + R_s)$ and
- to amplify the term $(R_p - R_s)$ at the cost of the term $(R_p + R_s)$ (Fig. 9-9d).
- The $(R_p - R_s)$-term is the anisotropic one containing the information about the monolayer conformation, whereas the $(R_p + R_s)$-term contains only the isotropic signals. The subsequent signal processing is performed as usual by fast Fourier-transformation (FFT) of the interferograms to obtain the spectra and their ratios (Fig. 9-9d) [11].

We have built up such a setup in our institute and are beginning now to investigate polypeptide monolayers directly *in situ* at the air/water interface with respect to
- the molecular conformation and organization directly at the interface
- conformation and organization changes due to the influence of external force fields (for example compression and dilatation of the layer), variation of the subphase, interaction with lipids, enzymes, special dyes etc.

9.6 References

1) G. Snatzke, Angew. Chem. **80** (1968) 15-26

2) G. Snatzke, Angew. Chem. **91** (1979) 380-393

3) A. Wollmer in W. Hoppe Biophysics 1983, p.144-151, Springer Verlag Berlin/New York

4) S. Brahms, J. Brahms, J. Mol. Biol. **138** (1980) 149-178

5) N. Greenfield, B. Davidson, G.D. Fasman, Biochemistry **6** (1967) 1630-1637

6) N. Greenfield, G.D. Fasman, Biochemistry **8** (1969) 4108-4116

7) L.L. France, J. Kieleczawa, J.J. Dunn, G. Hind, J.C. Sutherland, Biochim. Biophys. Acta **1120** (1992) 59-68

8) W.L. Zahler, D. Puett, S. Fleischer, Biochim. Biophys. Acta **255** (1972) 365-379

9) D.G. Cornell, J. Colloid Int. Sci. **70** (1979) 167-180

10) V. Erokhin, S. Vakula, C. Nicolini, Thin Solid Films **238** (1994) 88-94

11) T. Buffeteau, B. Desbat, J.M. Turlet, Applied Spectroscopy **45** (1991) 380-389

12) D. Blaudez, T. Buffeteau, J.C. Cornut, B. Desbat, N. Escafre, Applied Spectroscopy **47** (1993) 869-874

Address of the author:
Dr. Horst Hermel
Max-Planck-Institut für Kolloid- und Grenzflächenforschung
Rudower Chaussee 5
D-12489 Berlin-Adlershof

10 The Surface Structure of Lipid Drug Carriers - Influence on Carrier-Cell Interaction

Prof. Dr. L. Bergelson and Dr. A. Domb, Jerusalem

10.1 The Heterogenity of the Lipid Particle Surface

The surface structure of lipid particles[1] used as drug carriers influences numerous pharmacologically important carrier properties, such as stability, permeability, inter-membrane lipid exchange and particle homing as well as carrier-cell interactions. Of special significance is the fact that the surface structure of lipid particles is a regulator of lipase activity (Munderhwa and Brockman, 1992) and thus is an important factor determining the rate of drug release in the intestine and the circulation. A particularly important example is pancreatic phospholipase A2 which performs hydrolysis of lipids in the digestive system, and is acting selectivly only at interfaces between different lipid phases (Op den Kamp et al. 1975; Menashe et al. 1986). Formation of interfaces in lipid bilayers results also in enhanced permeability for cations (Papahadjopoulos et al. 1973; Cruzeiro-Hansson and Mouritsen 1988), water (Carruthers and Melchior 1983) and small organic molecules (Marsh et al.,1976). Obviously, knowledge of the carriers surface structure is essential for the design of efficient lipid-based drug delivery systems, and in order to understand why that or the other of these systems succeeds or fails as drug carrier. Most articles dealing with the pharmacology of lipid drug carriers provide a quite oversimplified view on their surface structure and neglect its important role in carrier function. Often the fluidity of the lipid surface is pictured as randomness and complete lateral disorder. However, an ever increasing amount of evidence has demonstrated that actually a certain degree of surface organization (lateral heterogeneity) is a fundamental consequence of the physical fact that the lipid carrier surface is a multimolecular system. This multimolecular character provides the lipid particle surface with a high degree of cooperativity, which is the main source of its lateral surface organization (Biltonen,1990).

Although increased surface rigidity generally prolongates the circulatory half-time of lipid carrier particles, the lipids commonly used for preparation of drug carriers are at physiological temperature entirely or partly in the fluid (liquid-crystalline) state. In

---

[1] In this paper the term 'lipid particle' is used to designate several lipid systems which can serve as drug carriers. These include lipid micelles, unilamellar or multilamellar vesicles (liposomes), phospholipid covered fat globulae (lipospheres), as well as serum lipoproteins.

principle such systems can undergo two types of cooperative processes: gelling, or crystallization which are associated with conformational ordering, and translational transition of the lipid molecules. These processes can occur simultaneously or separately. For proper understanding of the relationship between surface heterogeneity and lipid-carrier function it is important to distinguish between various levels of lipid lateral organization differing in their lifetimes and sizes. (Mouritsen and Jorgensen, 1995). In respect to time one can distinguish between "static" and "dynamic" heterogeneity. Static heterogeneity is the consequence of "solid-liquid" or "liquid-liquid" phase separation whereas dynamic heterogeneity is caused by thermal density fluctuations leading to the appearance of short lived gel-like microdomains in a fluid (liquid-crystalline) environment. The shape and position of these microdomains fluctuate. Nevertheless their appearance implies formation of low-ordered boundary regions which inspite of their short lifetime and small size serve as sites of ion penetration and enzyme reactions. They also may serve as nucleation centers preceding gross lipid phase separation induced by cations and are believed to play a significant role in the fusion of the lipid carriers with the plasma membrane of living cells. Considering sizes one can distinguish between macroscopic and microscopic heterogeneities although the borderline between them is somewhat obscure. On an operatonal basis the correspondig length scales may be defined as >500 nm and <1nm respectively. Intermediate length scales in the 1 - 100 nm range can be designated as mesoscopic.

A certain degree of dynamic heterogeneity is present even in lipid systems produced from an individual single acid phospholipid species. In practice, however, the lipids employed in the production of drug carriers usually are multicomponent mixtures of phospholipids with different fatty acid chains and/or different head groups and may include also glycolipids or cholesterol. At the surface of such multicomponent particles the coexistence of two or more lateral macrophases is possible. Sometimes it is assumed that the fast lateral diffusion of lipids in the fluid state results in deterioration of any structural patterns of the lipid surface. However, fluorescence microscopy has shown that even in relatively simple mixtures of lipids the lifetime of lateral heterogeneities or domains may be very long (see Kinnunen 1991). The actual lifetimes of the various micro- and macrodomains, as well as their size and shape, depend not only on the nature of the lipids but also on external factors such as pH, the presence and concentration in the surrounding medium of anorganic ions, proteins, peptides etc.

10.2 Methods to Study Lipid Surface Structure

In experimental studies of dynamic and static heterogeneity a broad range of physical metods have been employed including microscopy, calorimetry, ultrasound and various types of spectroscopy. The most direct way to visualize lateral lipid heterogeneity is by microscopy. To this extent various microscopic techniques such as freeze fracture and diffraction contrast electron microscopy can be used. Some recent developments in fluorescence digital imaging (Rogers and Glaser 1993) and differential polarization microscopy (Finzi et al. 1989) also offer sensitive ways for directly observing coexisting lipid domains. However, microscopy usually requires relatively large particles of cell size with fairly extended and stable domains. Differential scanning calorimetry (DSC), being a bulk equilibrium technique also provides only averaged information about the state of the lipid particle surface (see e.g. Wolf, 1992).

More information with less expenditure can be obtained by spectroscopic techniques.

In that connection it is highly important to understand the intrinsic time and distance scales to which the particular technique is sensitive. Thus, nuclear magnetic resonance, one of the most powerful techniques for studying the structure of lipid particles which provides valuable information on many of their features (e.g. head group orientation, chain order, transbilayer asymmetry, permeability and lateral diffusion) is characterized by a relatively long spectroscopic time scale. Although spin-lattice relaxation (T1) measurements are quite sensitive to fast motions, normally proton NMR is limited to time scales of several hundreds of microseconds and will not detect domains smaller than several hundred Angstroms.

With phosphorus NMR high resolution spectra can be obtained only for the smallest lipid particles. At the same time phosphorus NMR is a highly valuable tool in studies of lipid polymorphism, especially in detecting nonbilayer structures on the surface of lipid carriers (Gruner et al., 1985).

Much higher time and space resolution is characteristic for the methods of electron spin resonance (ESR) and fluorescence spectroscopy (time scales of less than ten nanoseconds). A main disadvantage of these methods is that they require introduction of extrinsic probes which inevitably perturb the host lipids and distribute randomely or with low specficity between different domains. To circumvent this obstacle many laboratories now are using as membrane probes modified natural phospholipids and glycolipids bearing a spin label or a fluorophore at the end of one of the apolar chains. In principle such probes are able to provide more reliable and detailed information because they report differentially on the physical state of microdomains enriched in their corresponding natural prototypes instead of showing averaged values for the

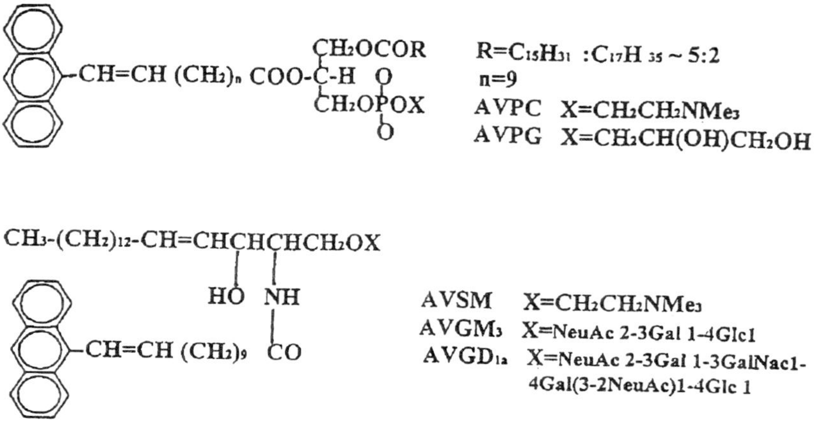

```
                    CH₂OCOR          R=C₁₅H₃₁ : C₁₇H₃₅ ~ 5:2
  ⬡-CH=CH(CH₂)ₙ COO-C-H  O           n=9
                    CH₂OPOX          AVPC  X=CH₂CH₂NMe₃
                         O           AVPG  X=CH₂CH(OH)CH₂OH
```

```
       CH₃-(CH₂)₁₂-CH=CHCHCHCH₂OX
                       |  |
                       HO NH          AVSM    X=CH₂CH₂NMe₃
                          |           AVGM₃   X=NeuAc 2-3Gal 1-4Glc1
   ⬡-CH=CH(CH₂)₉ CO                   AVGD₁ₐ  X=NeuAc 2-3Gal 1-3GalNac1-
                                              4Gal(3-2NeuAc)1-4Glc 1
```

Fig. 10-1: Anthrylvinyl-labeled phospholipids and glycolipids

whole surface. Unfortunately the spin and fluorescent labels commonly used in membrane studies are still of high perturbance even when attached to a lipid moiety. Thus, the nitroxide labels used in ESR are rather bulky and polar and have been shown to perturb significantly their lipid environment (Tailor and Smith, 1980). The same is true for most of the fluorophores commonly used for lipid labeling. Specifically, monolayer studies demonstrated that the polycyclic aromatic hydrocarbons pyrene and perylene are too bulky (Bredlow, et al. 1994), diphenylhexatriene labeled lipids are highly perturbing, because the fluorophore penetrates into the most tightly packed area of the lipid environment (Lentz, 1989), chains carrying a terminal dansyl or nitrobenzoxadiazol (NBD) group assume an unnatural configuration due to the polarity of the fluorophores (Molotkovsky et al., 1981; Chattopadhay and London, 1988), and with anthryloxy fatty acyl labeled lipids the fluorophore perturbs its neighbourhood strongly because it is linked perpendicularly to the acyl chain (Cadenhead et al.,1977). Therefore lipids labeled by any of these groups frequently prefer to distribute into fluid domains. Much less perturbation is caused by the flat and compact 2-anthryl group (Vincent et al.,1985), but that advantage is compromised by the low quantum yield requiring rather high probe concentrations.

A new family of fluorescent lipids. The multiprobe approach. These shortcomings can be largely overcome by labeling lipids with a 9-anthryl-vinyl (AV) group, a new fluorophore in membrane research (Fig. 10-1) which was chosen because it combines low perturbance (Bredlow et al., 1993) with favourable photophysical properties (Bergelson et al., 1985; Johansson et al., 1989). When attached to the chain at an appropriate distance from the polar headgroup, the AV-fluorophores reside exclusively in the center of the bilayer where the host lipids are packed most loosely, and affect neither the mobility nor the orientation of the polar head groups of the host lipids

(Molotkovsky et al., 1984). Significantly, the AV-labeled lipids disturb the packing of their unlabeled headgroup analogues to a much smaller extent than that of 'alien' species with the same fatty acids but with other headgroups (Bredlow et al., 1993). Accordingly, in heterogeneous multilipid systems they tend to distribute together with their unlabeled prototypes (Molotkovsky et al., 1991a,b; Gromova et al., 1992; Polozov et al., 1994) . In contrast to nonlipid probes the AV-labeled lipids sample only lateral heterogeneity and cannot sample vertically. Hence they are especially useful in studies of the lateral hetrogeneity of lipid surfaces induced by nonequivalent interaction of proteins with the various polar headgroups of different lipid classes (see below).

In order to minimize the ambiguities caused by probe perturbation it was suggested to use in parallel several AV-labeled lipids identical or closely related in all aspects esxcept the polar head groups (Bergelson et al. 1985). When located in the same environment such probes show close fluorescence parameters. However, in heterogeneous lipid matrixes these values may differ significantly if the probes partition differently between different domains. Thus, comparison of the probes' parameters can detect surface heterogeneities and even provide some information about the distribution of various lipid classes between different domains. If the probes respond differently to some stimulus, e.g. addition of cholesterol, insertion of a protein, addition of calcium etc., this indicates that a structural rearrangement of the host lipids must have taken place.

Multiprobe studies with AV-labeled lipid probes are conveniently performed by measuring head-group-dependent differences in the steady state polarization (or anisotropy) of their fluorescence. When a fluorophore in a viscous environment is excited with polarized light the fluorescence emitted is also polarized. The value of emission polarization is a measure of of the extent of the fluorophores motions during its excited lifetime (several nanoseconds for the AV-group). These motions are made up of two components: wobbling and rotational diffusion. Evaluation of the specific contribution of each component requires time resolved measurements depending on sophisticated expensive equipment. However for probing heterogeneity with AV-labeled lipids it suffices usually to compare the steady state polarization of several probes with appropriate headgroups in the same homogeneous environment and in the sample under investigation.

Of much importance in such comparative measurements is the fact that the AV-fluorophores have fixed orientations (due to the double bond) and allign uniformly and rigidly almost parallel to the long axis of the lipid molecules. Therefore the depolarization fluorescence practically depends solely on the rotational diffusion of the entire molecule which is of the same time scale as the fluorescence decay. Under these circumstances even subtle changes in the order of the probes environment are producing relatively big changes in the polarization of fluorescence. Some examples illustrating the power of the multiprobe approach are given below.

Popular models in physical studies of binary lipid systems are mixtures of phosphatidylcholine (PC) and sphingomyelin (SM), i.e. two phospholipids with the same headgroups and different apolar backbones, and mixtures of PC with phosphatidylethanolamine (PE) having different headgroups but the same fatty acids. The former system is also of practical interest in the design of lipid carriers because inclusion of SM into PC- (or PC-cholesterol) liposomes is an efficient way to increase their half-life in plasma (Gregoriadis and Senior, 1980). For mixtures of PC-SM with equal chain lengths it was concluded from calorimetric, X-ray and fluorescence polarization studies that the two components are freely miscible in the liquid- crystalline state (Calhoun and Shipley 1977; Lentz et al., 1981). That conclusion was challenged by results of fluorescence measurements using the multiprobe approach (Molotkovsky et al., 1991a). As can be seen from Fig. 10-2A, in a homogeneous matrix (individual dimyristoyl-PC) AV-labeled PC and SM showed very close fluorescence anisotropy (r) values, whereas in mixtures of fluid PC and gel state SM these values differed considerably. With increasing temperature the difference became smaller but it still persisted when both components were in the liquid-crystalline state demonstrating incomplete fluid-fluid miscibility of the host lipids.

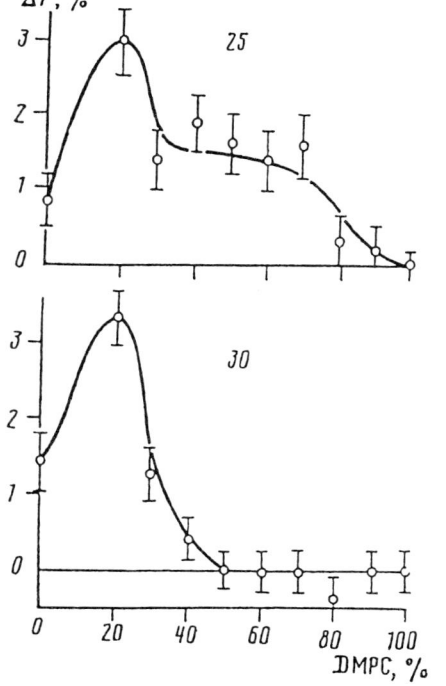

Fig. 10-2: Differences between the fluorescence anisotropy values of the anthrylvinyl-labeled phospholipids AVPC and AVSM in binary mixtures of "liquid" dimyristoyl-phosphatidylcholine (DMPC) and "solid" palmitoylsphingomyelin at different component ratios at 25 and 30 °C (Molotkovsky et al.,1991a).

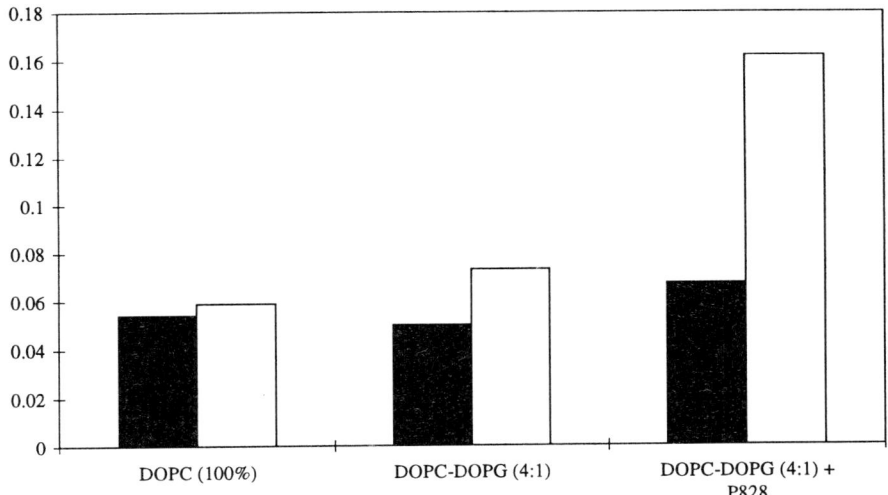

Fig. 10-3: Fluorescence anisotropy values of anthrylvinyl-labeled phosphatidylcholine (black bars) and phosphatidylglycerol (white bars) in pure dioleoylphosphatidylcholine (DOPC) and in a 4:1 mixture of DOPC and dioleoylphosphatidylglycerol (DOPG) in the absence and presence of the HIV envelope peptide P828 (Gawrisch et al.,1993).

A similar situation was found for mixtures of PC and PE with identical fatty acids. The phase diagrams for mixtures of dipalmitoyl-PC with dipalmitoyl-PE constructed on the basis of calorimetric, NMR and electron microscopic measurements indicated that in the fluid state these lipids form a single homogeneous phase (Luna and McConnell 1978; Petrov et al., 1982), although theoretical considerations and indirect data suggested that spontaneous segregation in the fluid state might be possible (Arnold et al., 1981; Blume et al., 1982). By contrast, the differences in the fluorescence anisotropy values shown by AVPC and AVSM in the same system demonstrated that their hosts demix even in the fluid state (Polozov et al., 1994). In the same way the existence of stuctural heterogeneity in fluid-fluid mixtures of PC with phosphatidylglycerol (PG) (Fig. 10-3) (Gawrisch et al., 1993) and of PC with phosphatidylserine (PS) (Gromova et al., 1992) was established. Although for the former system neutron scattering experiments indicated complete fluid-fluid miscibility (Knoll et al. 1992), the multiprobe approach revealed the existence of structural heterogeneity even when both components are in the liquid-crystalline state (Fig. 10-3).

Apparently in all cases mentionend the heterogeneities detected by the the multiprobe

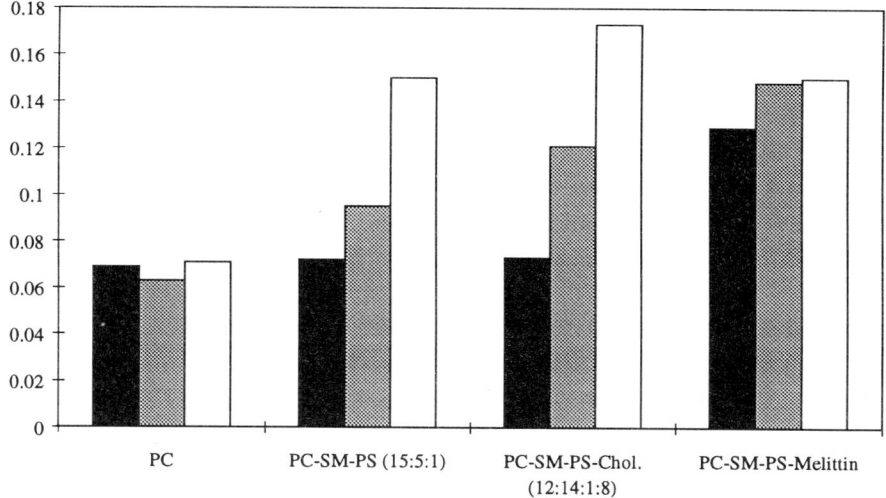

Fig. 10-4: Fluorescence anisotropy of AV -labeled phospholipid probes in egg phosphatidylcholine (PC) and in mixtures of PC, sphingomyelin (SM) and phosphatidylserine in the absence and presence of cholesterol (Chol) and melittin (Mel). Black bars, AVPC; gray bars, AVSM; white bars, AVPS (Gromova et al., 1992).

technique are of short range, dynamic nature, and the inconsistence with results obtained with other methods is caused by differences in the correlation times. Most probably the driving force of this type of lipid segregation is the formation of hydrogen bonds between molecules of SM, PE, PG or PS resulting in clustering of these species in an environment of fluid PC which is unable to participate in hydrogen bonding.

It is important to note that not every lipid species capable of hydrogen bonding separates from PC in binary mixtures. Application of the multiprobe approach to mixtures of glycolipids with PC using as probes AV-labeled cerebrosides and gangliosides demonstrated that, while cerebrosides tend to segregate from PC, the gangliosides appear to be completely miscible with PC even when their content in the mixture is as high as 15% (Molotkovsky et al., 1991b). Such behavior must be taken into account when using glycolipid-PC mixtures in carrier preparation, e.g. for the linkage of lipid carrier particles with proteins via Schiff bases, or for manipulating in vivo uptake profiles (see for example Allen and Chon 1987; Liu et al.,1992).

The examples cited so far referred to simple binary mixtures, however the multiprobe technique permits also to probe more complex systems. Thus for ternary mixtures of PC, SM and PS it was shown by multiprobing that PS concentrates in a separate phase

with relatively low lipid order even when its content does not exceed 5% (Fig. 10-4) (Gromova et al. 1992).
Noteworthy, in all lipid mixtures mentionend the mobilities of the different AV-labeled probes were equalized in the presence of 50% cholesterol (Fig. 10-4) indicating that at such concentration cholesterol eliminates the segregation of the parent phospholipids. As is well known at such level the sterol abolishes also phase transitions and suppresses interface dependent phenomena in single lipid systems. On the contrary, at low concentration cholesterol enhances lipid segregation in multilipid systems. Such dual effect opens interesting possibilities for manipulation of the pharmacological properties of lipid drug carriers.

10.3 Influence of Proteins and Peptides on the Surface Structure of Lipid Drug Carriers

The interaction of the carrier lipids with proteins and peptides is of practical importance because:
1) lipid carriers can be used for administration of proteins or peptide drugs;
2) after administration the carrier associates with endogeneous proteins and peptides (particularly, when liposomes are injected into the bloodstream their surface rapidly becomes coated with plasma proteins, and
3) proteoliposomes are of potential value for drug targeting.

As known from the textbooks, proteins can interact with lipids in two major ways: by associating (mainly electrostatically) with the lipid headgroups (peripheral proteins), or by embedding to various depth in the lipid bilayer (integral proteins) (for an overview see Watts and De Pont, 1985). Both types of proteins induce serious changes in the packing and lateral distribution of surrounding lipids. Although the protein-lipid interactions may not necessarily lead directly to large scale lipid segregation, formation of macrodomains may be facilitated by protein in lipid mixtures where the components have a disposition toward demixing. Both micro- and macroscale lipid reorganizations induced by proteins can easily be detected and partly characterized by the multiple probe technique, i.e. by comparing the behaviour of different AV-labeled lipid species in the corresponding lipid matrixes in the presence and absence of protein.
An example is the interaction of melittin, a bee venom protein which at high ionic strength penetrates into the lipid bilayer, with multicomponent liposomes made from mixtures of three phospholipids PC, PS and SM in the absence or presence of cholesterol (Gromova et al., 1992). As already mentionend these lipids are disposed

toward demixing in the absence of protein (Fig. 10-4). That trend increases markedly in the presence of melittin and appears to result in formation of a PS-enriched halo around the protein. As shown in Fig. 10-4, in PC-SM-PS vesicles doped with the three corresponding AV-lipid probes melittin caused an increase in the anisotropy of all three probes, however the PS probe was immobilized to a much higher extent than the corresponding analogues of PC and SM. Moreover, the transfer of fluorescence energy from the (single) tryptophan of melittin to the PS probe was noticeably higher than to the two other phospholipid probes. Such selectivity is not surprising and is consistent with earlier notions that the affinity of mellitin for acidic lipids is higher than for PC (see Tournois and de Kruiff, 1991, and the literature cited therein). However, in mixtures of PC with the stongly negative phosphatidylglycerol no significant preference of melittin for the acidic lipid was observable by the multiprobe technique (Fig. 10-4) (Gromova et al., 1992). It could be concluded that melittin displays very different affinities for acidic phospholipds with different headgroups.

Peripheral proteins usually have less pronounced effects on the mobility of the lipid chains than integral proteins, and the lipid species interacting directly with the protein are not always segregating from the bulk lipids. However several peripheral proteins and peptides are able to induce acidic domains in multilipid bilayers. One straightforward example is the interaction of phospholipids with a small basic peptide representing the C-terminal region of the envelope glycoprotein of human immunodeficiency virus (Gawrisch et a. 1993; Konig et al. 1995). This peptide designated as P828 contains several positively charged arginins. It does not interact with phosphatidylcholine but binds to negatively charged lipid surfaces. Multiprobe experiments performed with mixtures of phosphatidylglycerol and phosphatidylcholine showed that the peptide binds to such mixtures immobilizing selectively only the phosphatidylglycerol probe (Fig. 10-4). In a similar way large proteins containing clusters of basic aminoacids such as proteinkinase C can induce separation of acidic lipid species in multilipid mixtures.

10.4 Influence of Drug Loading on the Carrier Surface

Drug loading may significantly change the chain order and diffusion rates as well as the domain organization of the carrier surface, and these alterations may affect the the interaction of the carrier particle with plasma components and cells. Nevertheless the influence of drug loading on the lipid particle structure has still not be investigated systematically.

The group of drugs best studied in this respect are probably local anesthetics which are thought to act via interaction with membrane lipids (Trudell 1977). Different species of local anesthetics were shown to disturb the lipid surface to different extent (Singer and Jain, 1980). The degree of their interaction with PC was found to decrease in the order dibucaine>tetracaine>procaine. Dibucaine interacts most strongly with the polar head groups, procain interacts weakly at the region of the glycerol backbone, whereas tetracaine inserts within the acyl chains .

Some other drugs whose influence on the lipid carrier structure has been reported include non steroid antiinflammatory agents like indomethacin (Lasoner and Weringa 1990), the beta-blocker propanolol (Betageri et al., 1989), antibiotics (doxorubicin, antibiotic 537A) (reviewed by Fenske, 1993). Each of these drugs appears to modify the molecular organization of the carrier lipids, but the interaction mechanisms seem to be quite different. Thus, indomethacin enhances the lateral heterogeneity of the phospholipid membrane, propanolol displaces cations from the carriers surface, whereas doxorubicin has no effect on the headgroup conformation and permeability of PC vesicles, but induces cation permeability and fusion in vesicles containing besdes PC negatively charged phospholipids.

Remarkably up to now nothing is known with certainty about the effects of loading the lipid carries with these and other drugs on the behavior of the carrier particles in vivo.
Influence of the surface structure of lipid drug carries on their interaction with cells.
The interaction of lipid vesicles and related systems with cells can occur in several partly interrelated ways (Pagano and Weinstein, 1978; New et al., 1989):

- Physical absorption;
- Fusion;
- Endocytosis;
- Lipid transfer.

Each of these pathways depends specifically on the nature of the cells as well as on the surface structure of the lipid carrier.

Physical absorption

Physical absorption is believed to result in uptake of the drug carrier only if the carrier particle attaches to some of the cell surface proteins (Leserman et al., 1981; Margolis et al.,1982a). Appearantly phosphatidylcholine liposomes and low density lipoproteins (LDL)compete for the same cell-surface protein which is not the high affinnity LDL receptor (Galkina et al. 1991). Due to a still unknown mechanism absorption of phospholipid vesicles by cells causes a sharp increase in the permeability of the vesicle membrane (Margolis et al. 1982b), and water soluble drugs will leak out ("contact release"). This can lead to high local cocentration of the drug in the nearest environment of the cell, and under certain circumstances may result in intracellular accumulation of drugs.

Fusion

Fusion of lipid carriers with cells results in complete mixing of the carriers surface lipids with those of the plasma membrane of the cells without significant loss of the carriers content. In the case of multilamellar vesicles or lipospheres this may involve introduction of the internal lipid lamellae or the liposphere core into the cytoplasm. In vivo fusion of lipid particles with cells may be only of secondary significance because particles, if not protected by special coating (see below), are eliminated from the blood by phagocytic cells much faster then the time required for fusion.

Endocytosis.

During *endocytosis* lipid particles are first taken up into invaginations of the plasma membrane (endosomes) which then fuse with lysosomes where the drug carried by the lipid particle is released. If the carriers surface is associated with a ligand having a specific receptor on the cell surface, the carrier particles may be internalized via coated pits with subsequent recycling of the ligand.

Lipid transfer

Lipid transfer is the exchange of lipid molecules between the lipid carrier and cells. Exchange can occur spontaneously, and is significantly accelerated by exchange proteins. Such proteins are present in plasma, and appear to be involved in the exchange of phospholipids between serum lipoproteins and lipid carriers. In certain cases such protein-mediated lipid exchange can completely destroy the carriers membrane.

For a given cell type the prevalence of one or the other of these interaction mechanisms depends largely on the structure of the lipd carrier surface. With many cells absorption of phosphatidylcholine vesicles is greatest when the phospholipid is in the gel state, whereas fusion appears to be promoted when both gel-state and liquid-crystalline phosphatidylcholines are present on the particle surface. For example unilamellar liposomes composed of a mixture of "fluid" PC with gel-state SM fuse together at higher rate than than vesicles made from either fluid PC or gel-state SM. Fusion is also facilitated when the lipids on the carrier surface are adopting nonbilayer structures. In this context it is of interest to note that PC which in liposomes is always in the bilayer state exits in a nonbilayer form on the surface of small lipospheres (Domb et al.,1996).

Spontaneous lipid exchange occurs faster with carrier particles having a "fluid" lipid surface than with rigid ones. Rigidification of egg-PC vesicles by incorporation of cholesterol significantly reduces the exchange rate, and " rigid" sphingomyelin is less exchangeable than "fluid" phosphatidyl choline.

Our knowledge of the influence of the lipid surface structure on endocytosis is still very fragmentary. Generally, liposomes of low or intermediate fluidity are taken up by phagocytotic cells of the RES less avidly than more rigid vesicles. Incorporation into PC liposomes of the ganglioside GM1 inhibits phagocytosis (Allen and Chonn, 1987), whereas the ganglioside GM3 increases the uptake of lipid and lipoprotein particles by macrophages (Prokazova et al., 1991). Ceramides were reported to inhibit internalization of lipid particles by fibroblasts which must be taken into account when designing lipid drug carriers for topical application.

10.5 References

1. Allen, T.M, and Chonn, A. (1387) Large unilamcllar liposomes with low uptake into the reticuloendothelial system. FEBS Letters , 223, 42-44.

2. Arnold, K., Losche, A. and Gawrisch, K. (1981) ^{31}P NMR investigations of phase separation in phosghatidylcholinephosphatidylethanolamine mixtures. Biochim. Biophys. Acta 645, 143-148.

3. Razzi, M.D. and Nelsestuen,G.L. (1991) Extensive segregation of acidic phospholipids in membranes induced by proteinkinase C and related proteins. Biochemistry 30, 7961-7969.

4. Bergelson, L.D., Molotkovsky,J.G. and Manevich, E.M. (1985) Lipid specific probes in studies of biological membranes. Chem, Phys. Lipids, 37, 165-195.

5. Betagieri,C.V.,Thierault,Y. and Rogers, J.A. (1989) Correlation of partitioning of nitroimidazoles in the n-octanol/saline and liposome systems with pharmacokinetical parameters.Pharm. Res, 6, 399-403.

6. Biltonen, R.C. (1990) A statistical thermodynamic view of cooperative structural changes in phospholipid bilayer membranes:their potential role in biological function. J.Chem. Thermodynamics 22, 1-19.

7. Blume, A.R.J., Wittebort, S.R., Das Gupta and Griffin,R.G. (1982) Phase equilibria, molecular conformation and dynamics in phosphatidylcholine/phosphatidylethanolamine bilayers. Biochemistry 21, 6243-6253.

8. Bredlow, A., Galla, H.J. and Bergelson, L.D. (1993) Influence of fluorescent lipid probes on the packing of their environment. Chem.Phys, Lipids 62, 293-231.

9. Buser, C.A., Kim, J., McLaughlin, S. and Peitzsch, R.M. (1995) Does the binding of clusters of basic regidues to acidic lipids induce domain formation in membranes? Molec. Membrane Biol.12, 69-75.

10. Cadenhead, D.A., Kellner, B.M.J., Jacobson, K. and Papahadjopoulos, D. (1977) Fluorescence probes in model membranes. Anthroyl fatty acid derivatives in monolayers and liposomes of dipalmitoylphosphatidylcholine. Biochemistry 16, 5386-5392.

11. Calhoun, W.I. and Shipley, G.G. (1979) Sphingomyelin-lecithin bilayers and their interaction with cholesterol. Biochemistry, 18, 1717-1722.

12. Carruthers, A. and Melchior, D.L. (1983) Studies of the relationship between bilayer permeability and physical state. Biochcmistry 22, 5797-5787.

13. Chattopadhay, A. and London, E. (1988) Spectroscopic and ionization properties of N-(7-nitrobenz-2-oxa-1.3-diazol-4-yl)-labeled lipids in model membranes. Biochim. Biophys. Acta, 938, 24-34.

14. Cruzeiro-Hansson, L. and Mouritsen, O.G.(1988) Passive ion permeability of lipid membranes modelled via lipid-domain interfacial area. Biochim.Biophys. Acta 944, 63-72.

15. Domb A.J., Bergelson L.D. and Amselem S.(1996) Lipospheres for controlled delivery of substances. In: Benita (ed.) Microencapsulation, Marcel Dekker. pp, 377-410.

16. Fenske D.(1933) Structural and motional properties of vesicles as revealed by nuclear magnetic resonance. Chem.Phys. Lipids 64, 113-162.

17. Finzi,L., Bustamante,C., Garah, G.and Juang, C.B. (1989) Direct observation of large chiral domains in chloroplast thykaloid membranes by differential polarization microscopy. Proc.Natl. Acad.Sci. USA 86, 8748-8752.

18. Galkina, S.L., Ivanov, V.V., Preobrashensky, S.N., Margolis, L.B. and Bergelson, L.D. (1991) Low density lipoproteins interact with liposome binding sites on the cell surface. FEBS Letters 287, 19-22.

19. Gawrisch, K.,Han, K.H., Yang, J,S., Bergelson, L.D. and Ferretti, J.A. (1993) Interaction of peptide fragment 828-848 of the envelope glycoprotein of human immunodeficiency virus type I with lipid bilayers. Biochemistry 32, 3112-3118.

20. Genz, A. and Holzwart, J.F.(1986) Dynamic fluorescence measurements on the main phase transition of dipalmitoylphosphatidylcholine vesicles. Biophys. J. 50, 1043-1051.

21. Gregoriadis,G.and Senior,J.(1980) The Phospholipid component of small unilamellar liposomes controls the rate of clearance of entrapped solutes from the circulation. FEBS Letters 119, 43-46.

22. Gromova, I.A., Molotkovsky, J.G and Bergelson, L.D.(1992) Anthryl-vinyl-labeled phospholipids as membrane probes.The action of melittin on multilipid systems. Chem. Phys. Lipids 60, 235-246.

23. Gruner, S., Cullis, P.R., Hope, M.J.and Tilcock C.P.S. (1985) Lipid polymorphism. The molecular basis of non-bilayer phases. Ann. Rev. Biophys. Biophys. Chem.14, 214-238.

24. Hresko, R.C., Sugar, J.P., Barenholz, Y. and Thompson, T.E, (1987) The lateral distribution of pyrene-labeled sphingomyelin and glucosylceramide in phosphatidylcholine bilayers. Biophys. J. 51, 725-733.

25. Johansson, L.B.A., Molotkovsky, J.G. and Bergelson, L.D. (1989) Fluorescence properties of anthrylvinyl-labeled lipid probes. Chem. Phys. Lipids 53, 185-189.

26. Kinnunen,P.K.J.(1991) On the principles of functional ordering in biological membranes. Chem. Phys. Lipids 57, 375-399.

27. Knoll, W., Schmidt, G.,Rotzer, H., Henkel, T., Pfeiffer, W., Sackmann, E., Mittler-Neher, S. and Spinke, J. (1991) Lateral order in binary lipid alloys and its coupling to membrane function. Chem. Phys. Lipids 57, 363-374.

28. Konig, B.W.,Bergelson, L.D., Gawrisch, K. and Ferretti, J.A, (1995) Effect of the conformation of a peptide from gp41 on domain formation in model membranes. Molec. Membrane Biol.12, 77-82.

29. Lasoner, E. and Weringa, W.D. (1990) An NMR and DSC study of the interaction of phospholipids with some anti inflammatory agents.

30. Lentz, B.R. (1989) Membrane 'fluidity' as detected by diphenylhexatriene probes. Chem. Phys. Lipids 50, 171-190.

31.Lentz,B.R.,Hoechli,M. and Barenholz,Y. (1981) Acyl chain order and lateral domain formation in mixed phosphatidylcholine-sphingomyelin multilamellar and unilamellar vesicles. Biochemistry 20, 6803-6809.

32. Leserman, R.E., Barbet, J., Kourilsky, F. and Weinstein, J.F.(1980) Targeting to cells of fluorescent liposomes coupled with monclonal antibody or protein A. Nature, 288, 602.

33. Liu, D., Mori, A. and Huang, L. (1992) Role of liposome size and RES blockade in controlling biodistribution and tumor uptake of GM1- containing liposomes. Biochim. Biophys. Acta 1104, 95-101.

34. Luna, E.J. and McConnell, H. (1978) Multiple phase equilibria in binary mixtures of phospholipids. Biochim. Biophys. Acta 509, 195-204.

35. Margolis, L.B., Neyfakh, A.A.jr., Bergelson L.D. and Vasiliev, J.M. (1982a) Interaction of solid liposomes with epithelial cells. Cell. Biol. Int. Rep. 6, 131-136.

36. Margolis, L.B., Victorov, A.V. and Bergelson, L.D. (1982b) A novel mechanism of transfer of liposome-entrapped substances into cells. Biochim. Biophys. Acta 720, 259-263.

37. Marsh, D. (1995) Lipid-protein interactions and heterogeneous lipid distribution in membranes. Molec. Membrane Biol. 12, 59-64.

38. Marsh, D., Watts, A. and Knowles, P.F. (1976) Permeability of tempo-choline into dimyristoylphosphatidylcholine vesicles at the phase transition. Biochemistry 15, 3570-3578.

39. Menashe, M., Romero, G., Biltonen, R.I. and Lichtenberg, D.(1986) Hydrolysis of dipalmitoylphosphatidylcholine small unilamellar vesicles by pancreatic phospholipase A2. J. Biol. Chem. 261, 5328-5333.

40 . Molotkovsky , J . G., Dmitriev , P.J ., Molotkovskaya, I .M., Manevich , E.M. and Bergelson, L.D. (1981) 9-Anthrylvinyl-labeled phosphatidylcholine and sphingomyelin as fluorescent membrane probes. Bioorgan. Khim.7, 386-391.

41. Molotkovsky, J.G., Gromova, I.A. and Bergelson, L.D. (1991a) Component segregation in phosphatidylcholine-spingomyelin bilayers. Biol. Membrane 8, 934-943.

42. Molotkovsky, J.G., Imbs, A.B., Mikhalev, I.I. and Bergelson, L.D. (1991b) Molecular organization of glycosphingolipids in mixed composition lipid bilayers. Chem. Phys. Lipids 58, 767-774.

43. Molotkovsky, J.G., Manevich, E.M., Babak, V.I. and Bergelson, L.D. (1984) Perylenoyl- and anthrylvinyl-labeled lipids as membrane probes. Biochim. Biophys. Acta 778, 281-288.

44. Mouritsen, O.G. and Jorgensen, K. (1995) Micro, nano- and meso-scale heterogeneity of lipid bilayers and its influence on macroscopic membrane properties. Molec. Membrane Biol. 12, 15-20.

45. Munderwha, J.M. and Brockman, H.L. (1992) Lateral lipid distribution is a major regulator of lipase activity. J. Biol. Chem. 267. 24184-24192.

46. New, R.R.C., Black, C.D.V., Parker, R.J., Puri, A. and Scherphof, G.L. (1989) Liposomes in biological systems, in: "Liposomes, a practical approach" (ed. New, R.R.C.), IR Press, Oxford, pp. 221-257.

47. Op den Kamp, J.A.F., Kanertz, M.T. and Van Deenen, L.L.M. (1975) Action of pancreatic phospholipase A2 on phosphatidylcholine bilayers in different physical states. Biochim. Biophys. Acta 406, 169-177.

48. Pagano, R. and Weinstein, R. J. (1978) Liposome-cell interactions. Ann. Rev. Biophys. Bioeng. 7, 435-483.

49. Paphadjopoulos, D., Jacobsen, K., Nir, S. and Isac, T. (1973) Phase transitions in phospholipid vesicles. Fluorescence polarization and permeability measurements concerning the effect of temperature and cholesterol. Biochim. Biophys Acta 311, 330-338.

50. Petrov, A.G., Gawrisch, K., Brezesinski, G., Klose, G. and Mops, A. (1982). Optical detection of phase transitions in simple and mixed lipid-water phases. Biochim. Biophys. Acta 690, 1-7.

51. Polozov, I.V., Molotkovsky, J.G. and Bergelson, L.D. (1994) Study of phosphatidylcholine-phosphatidylethanolamine binary systems using lipid-specific fluorescent probes. Chem. Phys. Lipids 69, 209-214.

52. Prokazova, N.V., Mikhalenko, I.A. and Bergelson, L.D. (1991) Ganglioside GM3 stimulates the uptake and processing of low-density lipoproteins by macrophages. Biochem. Biophys. Res. Commun. 117, 582-557.

53. Rogers, R. and Glaser, M.(1993) Fluorescence microscopic imaging of membrane domains, in: "Optical Microscopy: emerging methods and application (eds. B.Herman, J.Lemasters) San Diego, Academic Press, pp.263-283.

54. Singer, S.J. and Jain, H.K. (1980) Interaction of four local anesthetics with phospholipid bilayer membranes. Can. J. Biochem. 58, 815-821.

55. Slepushkin, V.A., Starov, A.I., Bukrinskaya, A.G., Kogtev, L.S., Vodovozova, E.L., Timofeeva, N.G., Molotkovsky, J.G. and Bergelson, L.D. (1988) Interaction of influenza virus with gangliosides and liposomes containing gangliosides. Eur. J. Biochem. 173, 599-605.

56. Tailor, M.G. and Smith, J.C.P. (1980) The fidelity of response by nitroxide spin probes to changes in membrane organization. Biochim. Biophys. Acta. 945, 221-245.

57. Tournois, H., de Kruijff, B. (1991) Polymorphic phospholipid phase transitions as tools to understand peptide-lipid interactions. Chem. Phys. Lipids 57, 327-340.

58. Vincent, M., Gallay, de Bony, J. and Tocanne, J.F. (1985) Steady state and time resolved fluorescence anisotropy studies of phospholipid molecular motion in the gel phase using 1-palmitoyl-2-[9-(2-anthryl)-nonanoyl-sn-glycero-3-phosphocholine as probe. Eur. J. Biochem. 150, 341-347.

59. Watts, A.ed. (1993) Protein-lipid interactions. New Comprehensive Biochemistry, Vol. 25A, Elsevier, Amsterdam.

60. Wolf, D.E. Lipid domains: the parable of the blind men and the elephant. Comments Mol. Cell. Biophys. 8, 83-85.

Addresss of the authors:
Prof. Dr. Lev Bergelson and Dr. Abraham Domb
School of Pharmacy, The Hebrew University of Jerusalem
P.O.B. 12065
Jerusalem, 91120 Israel

11 Surface Area Analysis of Finely Divided and Porous Solids by Gas Adsorption Measurements

T. Schoofs, Coulter Electronics GmbH Krefeld

11.1 Introduction

Surface area of finely dived solids is mainly governed by the presence of pores in the solid matrix. A number of methods are available to characterise a solids surface area; depending on the degree of information required the user can choose a different approach. Physical adsorption of gases is a very popular method for the characterisation of mesoporous solids, materials often encountered as adsorbents, molecular sieves, catalysts and fillers. An overview of the available methods using the gas adsorption technique is given. The process of adsorption is described along with the different types of isotherms that may be encountered. The calculation of specific surface area is explained and finally a practical test was carried out on a typical sample showing the performance of modern day equipment.

11.2 Fundamental Understanding of Surface Area

In particle technology there are a number of cases where properties of the system are better described by the surface area of the particles than by their particle size. This is for instance the case when dealing with catalysts or pharmaceuticals. In the first case the exposed area is a measure of how fast the reaction at the surface will occur, whereas in the second example of the pharmaceuticals it is often difficult to find a suitable liquid for dispersion - these substances dissolve more or less in any liquid. As in the case of the catalysts, also here the action of the pharmaceutical is primarily linked to its exposed surface, i.e. resorption is controlled by the amount of area exposed to body-fluids. The problem in characterising surface area is trying to express a three-dimensional structure by a two-dimensional measure. A simple case is when subdividing a non-porous cube with edge length of 1 cm, having a surface of 6 cm^2, into smaller cubes with edge lengths of 1 mm (0.1 cm). We will then get 1000 smaller cubes with an exposed surface of 6 x 0.01 cm^2 the total surface being 60 cm^2 further subdivision into 10^{12} cubes with edge lengths of 1 μm (10^{-4} cm) will lead to a total surface area of 60.000 cm^2 or 6 m^2. If the cube weighs 1 gram, the specific surface area of the subdivided solid is 6 m^2/g. This shows that subdivision of a solid always leads to an increase of surface area. The presence of irregularities on the surface or even pores have a much more dramatic effect. The presence of pores in the above mentioned example could easily lead to surfaces 100 times bigger than those calculated. This shows the importance of the inner structure (microstructure) compared to the geometric dimensions of a solid. Imagine we have to characterise a sponge; we would have to spread out the whole structure - "iron" it into the form of a sheet - and then measure up length and width of this sheet and arrive at a two-dimensional size for the three-dimensional structure. In the case of the sponge this would seem a possible, although tedious, procedure. In the case of activated charcoals this is definitely impossible to perform. The particles are very small and the pores are of a very small size (in the nanometer range). Specific surface areas of these materials easily exceed values of 1500 m² per gram of solid. The presence of pores increases the surface area of a solid. The smaller they are, the more dramatic this effect will be. Surface irregularities and pores come in different forms, see Fig. 11-1. All indentations of this particle, except for the unaccessible closed pores can be measured by some form of analytical technique, e.g. by gasadsorption.

The pores from b. to g. are termed the internal structure, or microstructure of the material.

Besides in different shapes, pores come in different sizes.

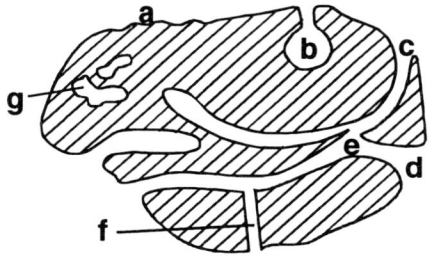

Fig. 11-1: Structure of a porous solid particle

      a.   surface roughness
      b.   ink-bottle shaped pore
      c. and f. cylindrical pores
      d.   funnel shaped pore
      e.   interconnecting pore
      g.   closed pore

Tab. 11-1: IUPAC classification of pores

	Porediameter
macropores :	d > 50 nm
mesopores :	50 nm > d > 2 nm
micropores :	2 nm > d

IUPAC, the International Union for Pure and Applied Chemistry (1) recommends the following classification (see Tab. 11-1).

11.3 Characterisation Methods

Several analytical techniques for characterisation of such porous systems are available (2). Dependent on the level of information we can choose from probing methods to detect surface-roughness - the surface level is probed with a diamond tip that measures the indentation-depths, sizing from about 2 µm up to 1 mm - up to gasadsorption measurements, detecting pores in the range of 0.6 nm to 200 nm. Tab. 11-2 shows an overview of the available methods.

Surface roughness

Roughness of (flat) surfaces can be determined by probing the surface with a diamond tip. The distance the probing - tip travels compared to the geometric dimensions of the sample surface is a measure of surface roughness. Only pores larger than about 2 µm and up to 1 mm may be detected.

Tab. 11-2: Overview of available methods for pore analysis with their application range.

Methods for pore analysis	Pore diameter (nm.)
Adsorption	10^0 – 10^2
Mercury- intrusion	10^1 – 10^6
X-ray diffraction	10^0 – 10^3
Electronmicroscopy	10^0 – 10^2
Permeametry	10^2 – 10^3
Lightmicroscopy	10^3 – 10^6
Surface roughness	10^3 – 10^6

Light microscopy

Microscopy and other techniques for size determination (e.g. sedimentation, electrical senzing zone, sieving and laserdiffraction) can provide surface area results by calculating the particle surface from the particles volume -which is measured in most cases-. One has to consider however, that all particle sizing methods will assume that the particles are smooth spheres. This will lead to results lower than those obtained by gasadsorption techniques, because surface irregularities and pores are not accounted for.

Permeametry

Air permeametry methods may be used to determine fineness of powdered samples. In these methods the powdered sample is compacted with a defined pressure and air is forced through the packed sample bed. The pressure differential is a measure of the sample surface. Again, also here the particles are considered to be non-porous. In some applications (e.g. cement and ceramics) particle size is derived from these experiments. The technique is known as Blaine test (3,4).

Electronmicroscopy

Like lightmicroscopy methods, electronmicroscopy can be used for pore analysis. Latter will allow detection of smaller pores because of the lower wavelength used. Both methods, however are limited to visible pores on the materials surface.

X-ray diffraction

X-ray diffraction uses the effect that pores create disturbances in the solid matrix. These disturbances will scatter light directed at the surface to a certain extent. The method should be referenced against a non-porous sample of the same material.

Mercury-intrusion

In mercury-porosimetry pores in the material are filled with mercury. Pressure is used to force the mercury into the pores of the material. The method is invasive, which means that after the pressure is relaxed, not all of the mercury will be expelled from the pores. Also employed pressure on the sample (up to 4.10^8 Pa) might cause it to collapse. The range of detectable pore sizes is extremely broad, ranging from 3 nm to 300 μm.

Another method, comparable to mercury intrusion is liquid porometry. Here the liquid is used to wet the pores completely - opposed to mercury-porosimetry, where the mercury does not wet the material under test-. In the subsequent analysis the liquid is forced out of the pores by increasing air-pressure. The measured pressure-flow relationship is a direct indication of the pore size distribution. The measurement range is from 0.05 µm to 300 µm.

Adsorption

Adsorption tests for evaluation of a materials exposed area can be subdivided in adsorption from liquids (solution) and adsorption from gases. Adsorption from liquids is sometimes used to determine surface area by the uptake of methylene-blue, iodine etc. This application is often used in the analysis of soots used as a filler in e.g. tyres. The most widely used method for characterisation of porous samples is the gasadsorption technique. Pores with diameters from 0.6 to 200 nm may be analysed.

11.4 Physical Adsorption of Gases

Surfaces of solids to be characterised by physical adsorption of gases must be cleaned of previously adsorbed (physisorbed) material. In most cases this will mainly be water, which is strongly adsorbed on most materials. To clean the surfaces, the samples are treated at elevated temperatures to speed up the cleaning proces. Either vacuum or flowing inert gas over the sample for some period of time will remove the adsorbed molecules. Temperatures and duration of the cleaning proces should be established empirically. The nature of the adsorbed molecules and the pore structure of the sample govern the time needed to effectively clean the sample. The temparature should be high enough to speed up the cleaning, without changing the sample physically or chemically in any way. The time needed may be established by checking the relation between time and measured result. After some preparation time one will observe no further increase in the resulting surface area. This combination of temperature and time period are the optimal outgas conditions.

Physical adsorption of nitrogen gas at low temperature (the boiling point of liquid nitrogen) is the most widely employed technique for the determination of surface area by gasadsorption.

Adsorption isotherms are usually constructed by measuring the uptake of gas (adsorptive) at increasing partial pressure over the sample (adsorbent). This is done in a step by step fashion in most volumetric and gravimetric experiments.

Surface area analysis requires that the adsorption isotherm, or at least part of it, is determined. Several techniques for recording these isotherms exist. Fig. 11-2 shows the schematic principles of the most common equipment. In volumetric methods (Fig. 11-2a) the amount of gas adsorbed at each consecutive step in the proces is derived from the residual gas pressure over the sample after pressure-equilibrium is achieved. Gravimetric instruments (Fig. 11-2b) determine the amount of gas adsorbed directly by weighing the mass increase of the sample with a microbalance. Theoritically, gravimetric techniques have some advantages over other techniques because the adsorbed gas amounts are detected directly. In practice however the equipment is very expensive, rather delicate and sensitive to vibrations. Carrier gas methods are sometimes used to determine surface area only (Fig. 11-2c). Here, a mixture of the adsorptive (N_2) and an inert carrier gas (usually helium) flows over the sample which is kept at liquid nitrogen temperature. The sample will adsorb part of the adsorptive from the mixture, hereby changing the heat conductivity of the gas mixture, which is measured with a thermal conductivity detector.

A practical drawback of this method is the effect of thermal diffusion on the measurement signal, distorting the adsorption signal. In practical experiments, the desorption signal is best analysed, because its peak is better defined. Obviously, this makes recording of a complete isotherms rather tedious.

Mainly practical reasons have lead to the popularity of the volumetric method for analysis of surface area and pore size distribution by gas adsorption. Over the last couple of years, microcomputer control of the measurement process and versatile data reduction made the technique faster and easier to use.

11.5 The Adsorption Process

Solid surfaces in contact with a gas (or liquid) will adsorb molecules. Forces controlling this process may be of chemical (chemisorption) or physical nature (physisorption).

Chemisorption occurs in one layer only and is e.g. used to determine a catalysts active metal area (also termed dispersion). The reaction is not readily reversible because the molecules react with the sample surface.

In physisorption the adsorbed gas molecules are held to the surface by weak (van der Waals) forces and the process is easily reversed by rising the sample temperature. An example of how this adsorption-desorption process progresses is shown in Fig. 11-3.

The interaction between the adsorptive and the adsorbent can take several forms, dependent on the interaction forces between solid and gas molecules and on the nature of the sample (porous or non-porous). Most isotherms encountered in practice fall into one of the categories shown in Fig. 11-4. The classification is based on work done by Brunauer, Deming, Deming and Teller (5).

The Adsorption Process

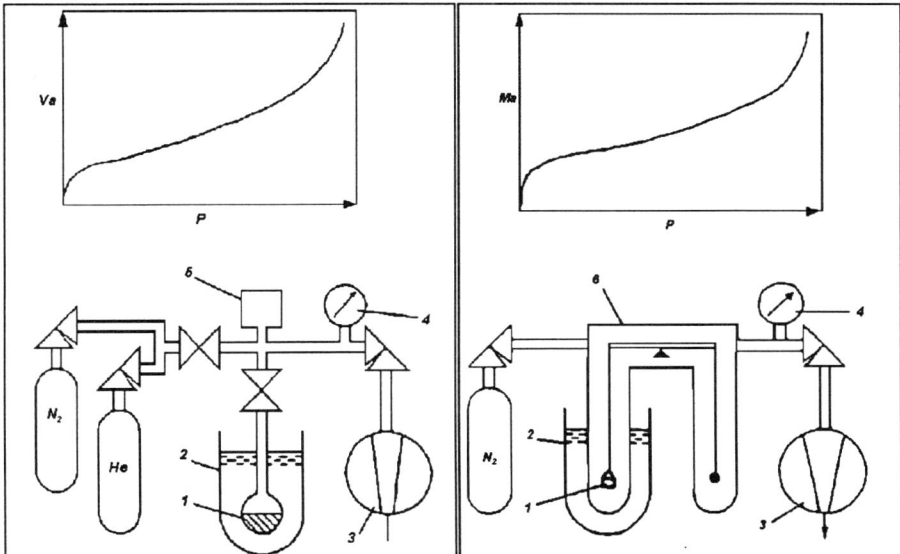

Fig. 11-2a: Volumetric method   Fig. 11-2b: Gravimetric method

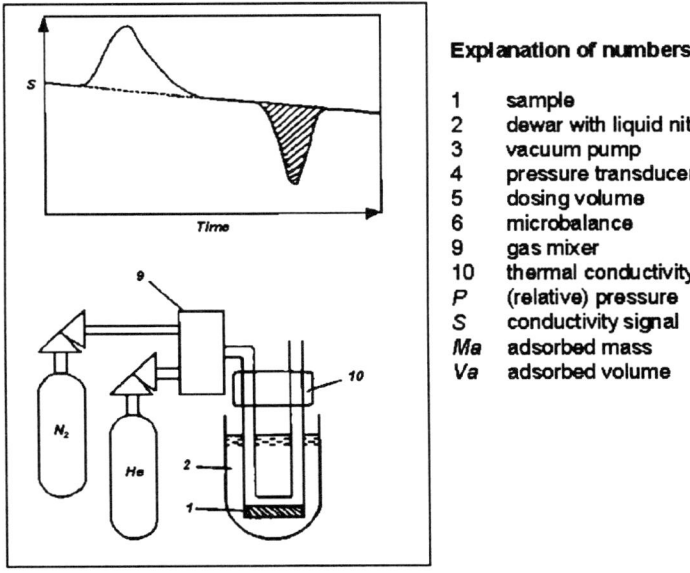

Explanation of numbers and symbols

1   sample
2   dewar with liquid nitrogen
3   vacuum pump
4   pressure transducer (manometer)
5   dosing volume
6   microbalance
9   gas mixer
10   thermal conductivity detector
P (relative) pressure
S conductivity signal
Ma adsorbed mass
Va adsorbed volume

Fig. 11-2c: Carrier gas method

Fig. 11-2: Shematic principles of usual equipment for gasadsorption experiments.

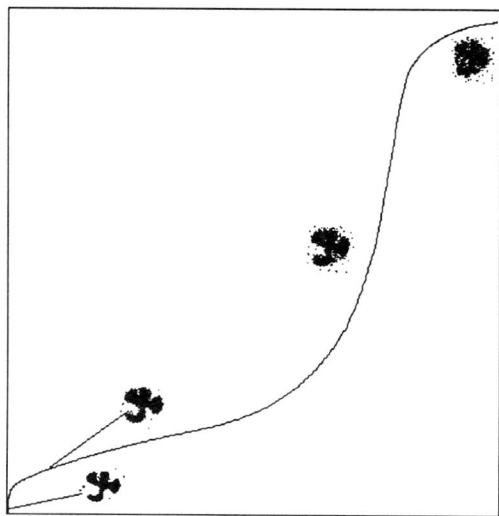

Fig. 11-3: Relationship between adsorbed layer build-up and development of the isotherm.

Type I isotherms are encountered with microporous materials, the steep rise at low p/p0 values indicating micropore filling. The plateau in the higher p/p0 region shows the absence of multilayer adsorption on the external surface of the sample which is often very small in microporous materials.

Type II isotherms are seen with nonporous or macroporous samples. Point B in the schematic represents the start of the linear section of the isotherm; considered to be the stage in the adsorption process where the monolayer is completed and multilayer adsorption is about to begin.

Type III isotherms are not very often seen. The convex curvature of the gradually increasing isotherm shows there is strong adsorbate-adsorbate interaction. these isotherm types have no distinct point B and do not allow B.E.T. surface area calculation.

Type IV isotherms are common for many mesoporous adsorbents. The hysteresis loop is typical for capillary condensation in mesopores. A type IV isotherm has its lower p/p0 range in common with the type II and hence shows a clear point B, allowing surface area calculation using B.E.T. theory.

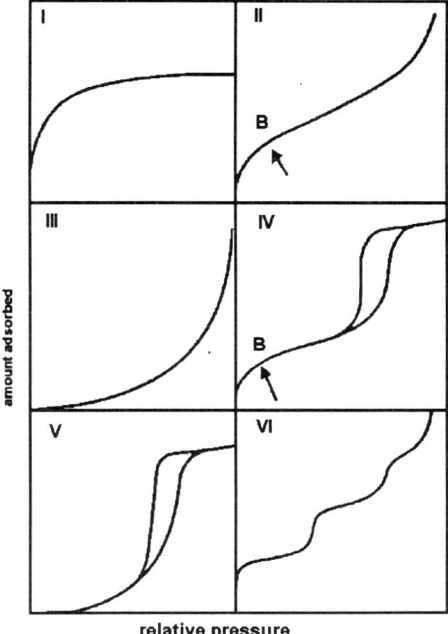

Fig. 11-4: Classification of physisorption isotherms

Type V may be compared to type III. The difference being the presence of mesopores. A few porous adsorbents show this adsorption behaviour, although it is uncommon.

Type VI is rare but may be seen with ideal multilayer adsorption on non porous surfaces. The step height represents the monolayer capacity for the adsorbed layers. Argon or Krypton adsorption at liquid nitrogen temperature on graphitized carbon shows this behaviour.

11.6 Determination of Surface Area

The generally used standard procedure for determination of the surface area of solids was initiated by Brunauer, Emmet and Teller (6) in 1938. Their theory, based on a simplified model proposes the use of the following equation to calculate the so called monolayer capacity from the amounts adsorbed at different relative pressures.

$$\frac{p}{n(p^0 - p)} = \frac{1}{n_m C} + \frac{(C-1)}{n_m C} \frac{p}{p^0}$$

n is the adsorbed gas amount at relative pressure p/p^0
n_m is the monolayer capacity
The factor C represents the B.E.T. constant and is related to the heat of adsorption (enthalpy) at the first adsorbed layer. It describes how strong the adsorbate molecules are attracted to the adsorbent surface. Its value - for surface area results to be reliable - should be between 20 and 200. A low number representing strong adsorbate-adsorbate interaction, leading to type III or V isotherms. C values > 200 indicate very intensive interaction between adsorbate and adsorbent. In those cases the interaction is so strong that not just monolayer-multilayer adsorption is occurring, but possibly micropore filling is taking place. This leads to isotherms of type I.
The form of the B.E.T. equation implies that there is a linear relationship between $p/n(p^0 - p)$ and p/p^0. This straight line is called the B.E.T. plot. It usually occurs within a range of 0,05 and 0,30 p/p^0. Solving the B.E.T. equation results in a value for n_m, the capacity of the monolayer.
Assuming a close-packed monolayer, a cross-sectional area for the nitrogen adsorbate molecule (a_m) of 0.162 nm^2 is generally accepted. The specific surface area can then be calculated:

$$A_{BET} = n_m \cdot L \cdot a_m$$

L represents Avogadro's constant.

The standard B.E.T. method requires 3 or more points in the relative pressure range of 0.05 to 0.30 p/p^0 to be taken.
In routine measurements a reduced technique is sometimes used. Here only a single point on the isotherm - usually in the range 0.25 to 0.30 p/p^0 - is recorded and the monolayer coverage is calculated by the reduced formula:

$$n_m = n(1 - p/p^0)$$

This simplification of the B.E.T. equation is only allowed if the value for C is sufficiently high, say > 100. If there is any doubt, the single point analysis should be calibrated against standard B.E.T. measurements. Another possible way to proceed is the use of reference samples of the same nature as the sample material (7).
When reporting B.E.T. surface area it is good practice to state the outgassing conditions (temperature, time), the linearity range for the B.E.T. plot, as well as the values for p^0, n_m and C. An example is shown in Fig. 11-5.

The analysis of very small surface areas (< 1 m²/g.) often requires the use of adsorbate gases other than N_2. Because the adsorbed gas amounts using nitrogen are minimal, the differences in pressure are too small to be resolved. Also in volumetric techniques, the size of the so called dead space plays an important role; the smaller it is, the more sensitive the detection of small pressure changes will be. In volumetric measurements the size of the dead space - the volume of the sample holder and the associated tubing that connects it to the dosing chamber - plays an important role with respect to resolution and should be recorded. This is performed using helium at the temperature of the subsequent adsorption test, because it does not adsorb on most solids. A possible way to get around the problem of the small pressure changes is the use of adsorbate gases with lower saturation vapour pressure (p^0). Noble gases like xenon or krypton are often used to size low surface areas. As an example, the saturation vapour pressure of krypton, a supercooled liquid at liquid nitrogen temperature, is only 2.5 mm Hg. In comparison, N_2 has a saturation pressure of around 760 mm Hg at the temperature of liquid N_2 (77 K). Provided the pressure sensing system is sensitive enough to measure these small pressures, it is possible to analyse specific surface areas as small as 0.01 m²/g.

11.7 Practical Measurement

A practical experiment was done using a B.E.T. reference material obtained from N.I.S.T. (National Institute for Standards and Technology, USA). The sample was an alumina (Al_2O_3) with a certified B.E.T. surface area between 156.2 and 161.2 m²/g. The instrument used in this test was the COULTER SA 3100, a recently introduced standard volumetric gasadsorption analyser

The material was weighed into a sample holder - approximately 0.1 g was taken - and evacuated for 3 hours at a temperature of 300°C.

The dry sample weight was recorded after this preparation and the sample was transferred to the analysis port of the instrument. A dead space evaluation is made prior to the adsorption test using N_2 as adsorptive.

A complete adsorption-desorption isotherm was measured on this sample in not quite 3.5 hours. A surface area only result will be available in under 20 minutes.

The reported specific surface area was 156.89 m²/g. The adsorption isotherm of the alumina is of type IV, (C-Value: 108). All other relevant parameters of the B.E.T. equation are stated in the report; goodness of fit, monolayer capacity and the relativepressures along with the respective adsorbed amounts of gas. In the instrument used in this test, the saturation vapour pressure is continuously monitored by a vapour

```
COULTER                              SA 3100
Serial No.      w48040       Software Version        1.05

Sample ID       NBSAL1       Start Date          01/31/95
Customer        COULTER      Start Time          05:54:12
Operator        TONY PEREZ   Elapsed time     3 hrs 23 min
Sample Weight   0.1115 g     Outgas Time         180 mins
Profile         FULLUN       Outgas Temperature     300 C

Surface Area Report
    BET Surface area          156.89 sq.m/g

    Slope                     0.2466
    Intercept                 0.00230
    C_value                   108.134
    Monolayer Volume          36.0363 cc/g (STP)
    Correlation coefficient   1.0000

    One Point BET Surface Area  (Ps/Po=0.3)   153.57 sq.m/g

Analysis Data
    Ps/Po          BET Function      Vads  cc/g(STP)
    0.0559         0.01600           33.223
    0.0662         0.01860           34.184
    0.0734         0.02041           34.811
    0.0815         0.02243           35.475
    0.0895         0.02442           36.104
    0.1001         0.02703           36.894
    0.1197         0.03186           38.288
    0.1401         0.03686           39.656
    0.1602         0.04178           40.948
    0.1799         0.04662           42.188
    0.1998         0.05156           43.434
```

Fig. 11-5: Example of B.E.T. analysis result print out.

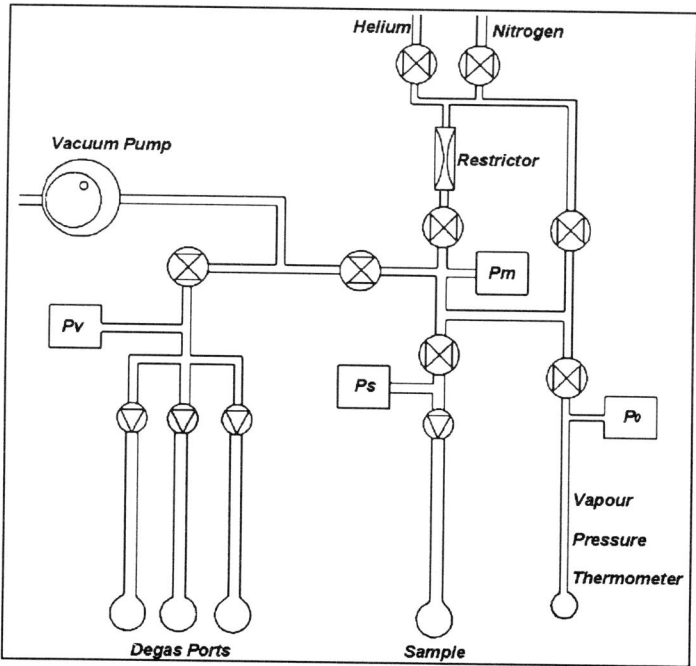

Fig. 11-6: Schematic diagram of volumetric gasadsorption analyser (COULTER SA 3100)

pressure thermometer (see Fig. 11-6). The instrument achieves very rapid results for a number of reasons; its dead space is kept to a minimum and its dosing system volume - the tubing section associated with the pressure transducer Pm in Fig. 11-6 - is small compared to other equipment available. The equilibration of the dosed gas quantities is monitored with a different transducer (Ps in Fig. 11-6), so the dosing manifold can be filled with the quantity of gas required for the next dosing step. This technique is called "concurrent dosing" and speeds up analysis remarkably.

11.8 References

1. IUPAC Recommendations 1984: Sing, K.S.W., Everett, D.H., Haul, R.A.W., Moscou, L., Pierotti, R.A., Rouquérol, J., Siemieniewska, T., "Reporting Physisorption Data for Gas/Solid Systems with Special Reference to the Determination of Surface Area and Porosity", Pure and Appl. Chem. 57 (1985) 4, 603-619.

2. Robens, E.: "Die Bestimmung der Spezifischen Oberfläche von Feinteiligen und Porösen Stoffen". TIZ Fachberichte, vol 110, No.3 (1986) 171-175.

3. DIN66126: Bestimmung der spezifischen Oberfläche pulverförmiger Stoffe mit Durchströmungsverfahren. *Grundlagen / laminarer Bereich.*

4. DIN66127: Bestimmung der spezifischen Oberfläche pulverförmiger Stoffe mit Durchströmungsverfahren. *Verfahren und Gerät nach Blaine.*

5. Brunauer, S. Deming, L..S., Deming, W.S., Teller, E. J. Amer.Chem.Soc.62 (1940) 1723

6. Brunauer, S., Emmett, P.H., Teller, E., J. Amer. Chem. Soc.60 (1938) 309.

7. Robens, E., Müller, U., Unger, K.K.: "Normung, Referenzmaterial und Ringversuche zur Oberfächen- und Porenbestimmung". Vakuum-Technik, 38.Jahrgang, Heft 3-4/1989.

Address of the author:
Ton Schoofs
Coulter Electronics GmbH
Europark Fichtenhain B 13
D-47807 Krefeld

12 Determinations of Surface Area in Comparison

H. Winter, Fachhochschule Nürnberg

12.1 Sample Preparation - Single-point- and B.E.T. - Multi-point Measurement

At research and application there often appear problems necessitating the knowledge of the surface areas of disperse materials, e.g. at the catalysis, the burning or drying. Besides this, the particle technology uses the specific surface area to determine the fineness of disperse materials by its integral mean.
The usually applied methods of measuring surface areas are:
A. flow methods
B. photometric methods
C. methods of sorption (physisorption).
With the methods A. and B. surface areas are measured which are resulting in its geometrical dimension; with method C. - the adsorption - the inner surface area can be measured additionally.
As the part of the inner surface area of high porous materials is essentially higher (e.g. catalysts) than the part of the outer surface area, only methods of sorption are suitable for determining the surface area of these materials.
There are two standardized methods to determine the specific surface area by gas adsorption:
1. The B.E.T. -multi-point measurement DIN 66 131 [1]
2. The single-point measurement according Haul and Dümbgen DIN 66 132 [1]
During June and July 1995 a Coulter® SA 3100, an automatically compact system to determine the surface areas according B.E.T. measurements, was available at the laboratory of Prof. Stieß, the department for mechanical process enigineering of the Georg-Simon-Ohm-Fachhochschule. During the practical training of the students this system was used for measurements in comparison with the Areameter from Ströhlein (single-point measurement according DIN 66 132). 22 measurements in 6 test rows and with 5 different materials were carried out. A description for better understanding follows under item 12.5 in which the results are explained.

12.2 Preparation of the Samples

Usually the quantity of materials for the analysis is higher than the quantity of samples necessary for the measurements. To obtain a representative sample it is important to make first a division of the sample (e.g. by means of a rotating sample divider).

12.2.1 Drying

Before the measurement, adsorbed impurities like steam have to be removed from the sample. For that, the divided samples were dried in a drier for approximately 24 hours at temperatures from 100 - 150°C, subsequently cooled down in an exsikkator partly filled with drying agent and stored until analyzing. This method refers to inorganic materials like the following. Organic materials often need temperatures > 50°C.

12.2.2 Outgassing

Besides drying a careful outgassing of the samples is important for exact measurements at the adsorption method. For the measurements mentioned in this article following methods for the outgassing were used:
- SA 3100 : heating up of the samples to 180°C at simultaneous evacuation
- Areameter : heating up of the samples to 180°C in an external outgas operation at simultaneous rinsing of the samples with analysing gas (N2).

To keep the samples' preparation time efficient, it is recommendable to determine the influence of the outgas time of the material and the outgas temperature in a test row. Herewith influences like sintering of the sample caused by a too long outgas time at temperatures too high could be noticed.

12.2.3 Quantity - Accuracy of the Quantity

The quantity of the samples should be chosen as large as possible. The limit results in:
- the minimum sufficient change in pressure in the system after adsorption has taken place
- the maximum quantity in the tube.

Usually the manufacturer recommends the optimum quantity.
Example for the SA 3100

Tab. 12-1

Estimated Specific Surface Area (m^2/g)	Mass of Sample for Analysis
>100	100mg
30 - 100	330mg
10-30	1,000g
3-10	3,330g
1-3	10g

As can be depicted from Tab. 12-1, for surface areas > $30 m^2/g$ only fractions of a gram are necessary. At small amounts weighing errors are especially significant. Possible sources of errors are:
- electrostatical charge
- impurity of the tube e,g. with sweat, dirt or fingerprints

In a test the weighing error caused by electrostatical charge was determined in a glass tube. The mass difference between the loaded and the unloaded tube was 0.0130 g. The tube was cleaned with a tissue before the weighing. At a sample weight of 0.2000 g, 0.013 g correspond to 6.5 %.

12.3 Choosing of the Adsorbate Gas

For surface area measurements usually nitrogen of a purity of 99.99 % is used as measuring gas. Other gases, especially inert gases must have following features:
- the gas have to be inert at the temperature measuring point
- the gas must have a nearly spherical shape
- with microporous samples the diameter of the molecules of the gas must be smaller than the pore diameter of the sample.

12.4 Conclusions

To get comparable measurement results the manufacturer of the product (product information) and the customer (measurements for the purpose of controlling goods received) have to agree upon equal conditions concerning the preparation of the measurements and the samples. If such an agreement is missing, the measurement results could deviate.

12.5 Description of the Measurement Results

Tab. 12-2: Overview upon the measurements

Measurement No	test row:	material:	object:
1 - 6	1	activated coal	influence of the outgassing
7 - 10	2	activated coal	influence of the quantity
11 - 15	3	catalyst K306	influence of the quantity at a catalyst with higher surface area
16 + 17	4	catalyst Ka	comparison with existing results
18 + 19	5	Zeolon 500	measurement of a micro-porous sample
20 - 22	6	ZnO (zinc oxide)	measurement of a comminuted product

Tab. 12-3: **Influence of the outgassing, test 1 - 6**
Results: "Surface Area Measurements in Comparison - Coulter SA 3100"
Object: Influence of a longer period of outgassing on the measurement results
Results: No significant change of the measurement results in dependence on the the preparation of the sample.

test no	1	2	3	4	5	6
sample no :	1	2	2	3	3	4
date :	31.05.1995	07.06.	07.06.	08.06.	08.06.	13.06.
kind of measurement :	BETNORM*	BETNORM	BETNORM	BETNORM	FULLUN*	BETNORM
Material :	activated coal	activated coal	activated coal	activated coal	activated coal	activated coal
desired quantity (gram)	0.3300	0.3300	0.3300	0.3300	0.3300	0.3300
actual quantity (gram)	0.3000	0.2997	0.2997	0.3375	0.3375	0.3450
drying : outgassing:	24h/125°C 2h/180°C	valid Exsikkator	for Exsikkator	all 2h/180°C	samples Exsikkator	mentioned Exsikkator
Sm /m^2/g = product info	78,5 m^2/g + - 2m^2/g	dito	dito	dito	dito	dito
Sm/m^2/g = SA3100	72.239	72.284	72.369	72.488	72.830	72.478
Sm (1 Pkt.) /m^2/g=SA3100	71.181	71.144	71.207	71.347	_	71.403
Sm /m^2/g DIN 66132 (Areameter)	74.96			75.32	75.32	74.79
measurement time (min)	51	27	27	50	306	29

*BETNORM = measurement in the relative pressure range P_s/P_0, 0.05 to 0.2.
*FULLUN = measurement of the complete adsorption isotherm

Tab. 12-4: Influence of the quantity - test. 7 - 10

Results: "Suface Area Measurements in Comparison - Coulter SA 3100"
Object: Influence of the quantity on the measurement results
Results: If about 1/3 of the recommended quantity is used the measured surface area is lower than the the comparable results of measurements 1 - 8.

test no	7	8	9	10
sample no :	5	6	7	8
date :	07.07.1995	07.07.	10.07.	10.07.
kind of measurement :	BETNORM	BETNORM	BETNORM	FULLUN
Material :	activated coal	activated coal	activated coal	activated coal
desired quantity (gram)	0.3300	0.3300	0.3300	0.3300
actual quantity (gram)	0.1318	0.3952	0.1016	0.1023
drying : outgassing:	24h/180°C 2h/180°C	valid 2h/180°C	for all 0.25h/180 °C	samples 1h/180°C
Sm /m²/g = product info	78.5 m²/g + - 2m²/g	dito	dito	dito
Sm/m²/g = SA3100	72.499	72.564	71.332	71.144
Sm (1 Pkt.) /m²/g=SA3100	71,382	71,421	70,01	69,841
Sm /m²/g DIN 66132 (Areameter)	74,9	dito	dito	dito
measurement time (min)	23	31	47	205

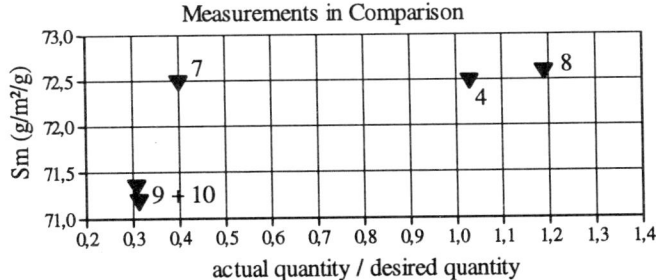

Fig. 12-1: Measurements no. 4 and 7 - 10 with activated coal - influence of the quantity

Tab. 12-5: **Influence of the quantity at the catalyst (Sm ca. 225 m^2/g)- test no 11 - 15**
Results: "Suface Area Measurements in Comparison - Coulter SA 3100"
Object: Influence of the quantity on the results of a material with a higher surface area
Result: No significant influence can be noticed.

test no	11	12	13	14	15
sample no :	10	11	12	11	13
date :	29.06.1995	30.06.	30.06.	11.07.	
kind of measurement :	BETNORM	BETNORM	BETNORM	BETNORM	FULLUN
Material :	Katalysator K 306	K 306	K 306	K 306	K 306
desired quantity (gram)	0.100	0.100	0.100	0.100	0.100
actual quantity (gram)	0.0613	0.096	0.1511	0.095	0.1053
drying : outgassing:	24h/125°C 1.5h/180°C	valid 1.5h/180°C	for 1.5h/180°C	all +1.5h/180°C	samples Exsi.+ 0.5/180°
Sm /m^2/g = product info	no information				
Sm/m^2/g = SA3100	225.39	229.98	224.35	229.42	217.55
Sm (1 Pkt.) /m^2/g=SA3100	220.13	224.70	219.12	223.88	212.12
Sm /m^2/g DIN 66132 (Areameter)	215.1				
measurement time (min)	40	40	44	41	301

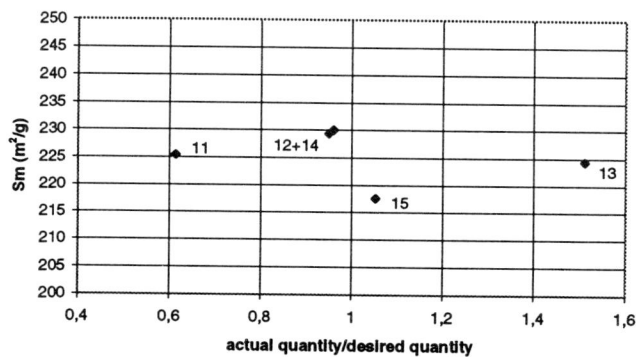

Fig. 12-2: Catalyst K 306 - test 11 - 15 influence of the quantity

Tab. 12-6: test no 16 + 17 - catalyst - as well as 18 + 19

Results: "Suface Area Measurements in Comparison - Coulter SA 3100"

Object: Comparison of the results of the surface area measured with the SA 3100 with the results from the practical lessons 1989-1995.

Results: The results of the measured surface areas are nearly corresponding with the tests mentioned above.

We have the results of 9 measurements in the range of Sm = 154,78 m²/g to 181,89 m²/g. The average value is 170,76 m²/g.

test no	16	17			18	19
sample no :	14	15			16	17
date :	13.06.1995	21.06.			21.06.	22.06.
kind of measurement :	BETNORM	BETNORM			BETNORM	BETNORM
Material :	catalyst	catalyst			Zeolon 500	Zeolon 500
desired quantity (gram)	0.1	0.1			0.1	0.1
actual quantity (gram)	0.114	0.1071			0.116	0.106
drying : outgassing:	24h/180°C	valid 2h/180°C	for Exsikkator	all	samples Exsikkator	1h/180°C
Sm /m²/g = product info	no information				400-440	400-440
Sm/m²/g = SA3100	182,14	169,06			260.4	248.96
Sm (1 Pkt.) /m²/g=SA3100	177.7	164.94			266.64	254.43
Sm /m²/g DIN 66132 (Areameter)	180				170	190
measurement time (min)	41	37			41	39

12.6 Test 18 and 19

Material: ZEOLON 500

Zeolon 500 is used as molecular filter with a pore size of 0.4 - 0.5 nm and a surface area of 400 - 440 m2/g (product information)

This product information cannot be confirmed by the measurement method used. The reasons for this could be:
- no physisorbtion according adsorption isotherm type 2 or 4 (this can be checked by means of the plot of the SA 3100 adsorption isotherm.)
- the proportion of the pores to the molecule diameter of the adsorbate gas (pore size = 0.4 - 0.5 nm; Surface requirements of the N2 molecule = 0.162 nm2).
- Outgas time is too low - with Zeolites several hours are necessary.

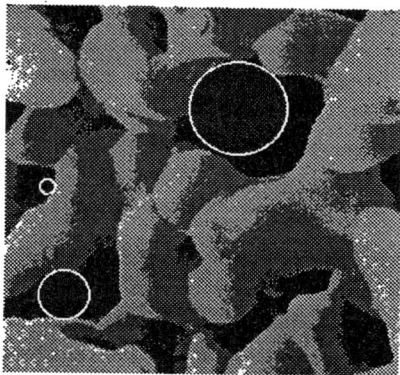

Fig. 12-3: Model of pore size proportion to molecule diameter

Only the small molecule depicted on the left side of the picture is able to cover all free areas of the pores.

The surface areas measured with the SA 3100 are significantly above the results of the Areameter. The reason for this could be that at the preparation of the sample for the measurement with the Areameter without vacuum, heated nitrogeneum was used for rinsing, and by doing this, a part adsorbtion happened.

Tab. 12-7: **test no 20 - 22**

    Results: "Suface Area Measurements in Comparison - Coulter SA 3100"
    Object: Measurement of a material without any inner surface area
    The quantity desired cannot be reached because of the too low volume of the tube.
    Result: Good agreement with the results measured with the Areameter.
    We additionally computed for test no 19 a surface area of 3.17 m²/g from a measured particle size distribution (Coulter ®-LS 130).

test no	20	21	22
sample no :	18	19	19
date :	22.06.1995	28.06.	28.06.
kind of measurement :	BETNORM	BETNORM	BETNORM
Material :	ZnO no.1	ZnO no.2	ZnO no.2
desired quantity (gram)	3.33	3.33	3.33
actual quantity (gram)	1.56	2.6125	2.6125
drying : outgassing:	24h/125°C 40min/180°C	dito dito	dito 2nd measurement
Sm /m^2/g = product info	no information		
Sm/m^2/g = SA3100	6.794	3.422	3.433
Sm (1 Pkt.)/ m^2/g=SA3100	6.58	3.366	3.377
Sm /m^2/g DIN 66132 (Areameter)	5.6	3.4	3.4
measurement time (min)	26	25	24

12.7 Summary

With the Coulter SA 3100 multi-point measurement according DIN 66 131 (B.E.T.-measurement) is possible.

The measurement results are well reproducibly comparable to the results determined with the Areameter.

The expenditure of time for a measurement in the scope of BET (approx. 10 measurement points) is 30 - 40 minutes and corresponds to the time for a measurement with the Areameter.

The SA 3100 operates automatically, so, mistakes made by the operator are avoided.

Besides the measurement results you can see its course from the plot of the adsorption isotherm.

The handling is without any problems and easy to learn.

12.8 References

[1] DIN - Taschenbuch 133: Partikelmeß-technik; Normen, 2. edition, Berlin/ Köln: Beuth; Wiesbaden/Berlin Bauverlag 1987

Address of the Author:
Herbert Winter
Wassertorstr. 10
D-90489 Nürnberg

13 The Surface Area of Magnesium Stearate: An Example for a Complex Analysis Task

Dr. F. Metz, Micromeritics GmbH

13.1 Introduction

Gas adsorption surface area values can predict the reaction rate of compounds such as medicinal products in tablet or powder form. The physical properties of components such as magnesium stearate, for example, influence the tableting characteristics of the formulation as well as the *in vivo* degradation and dissolution of the solid drug. Tablets and powders generally have low gas adsorption surface areas compared to other sample types, such as activated carbon or molecular sieves. Analysing the low surface areas with the accuracy and reproducibility needed for the precise prediction of the properties of a given formulation therefore needs increasing attention. The United States Pharmacopeia, which has already established a standard for the particle size distribution of magnesium stearate, may adopt surface area standards for this material in the near future.

13.2 Considerations for Small Surface Areas

In the widely used and well known static-volumetric technique for gas adsorption measurements the amount adsorbed by the sample is not measured directly. It is calculated from the difference of the volume dosed onto the sample and the one remaining non-adsorbed in the sample tube. With low surface area materials in the system nitrogen at liquid nitrogen temperature (77 K), due to the low adsorption taking place, these numbers are very similar, and the error introduced with the subtraction becomes significant. Assume a sample having a total surface area of 1 m^2, in a sample tube with a volume of 20 cm^3. At a relative pressure of $P/P_0 = 0.2$, only 0.287 cm^3 of gas will be adsorbed while 14,144 cm^3 will stay in the sample tube. The use of argon or krypton rather than nitrogen for the adsorbate improves the accuracy of low surface area measurements (see Tab. 13-1): using argon improves the accuracy by a factor of three, krypton gives a factor of 200.

Tab. 13-1: Effect of the adsorptive on the accuracy of the analysis

	N_2	Ar	Kr
P_0 @ 77 K / mm Hg	760	250	2,5
σ / nm^2	0,162	0,138	0,210
V_{ads} / cm^3 STP	0,162	0,138	0,210
V_{rest} / cm^3 STP	14,144	4,653	0,046
V_{ads}/V_{rest}	0,020	0,072	4,826
Rel. to N_2		3x	200x

Therefore, with krypton as adsorptive it is possible to analyse surface areas of as little as 10 cm^2/g (0.0001 m^2/g). One has to bear in mind, however, that the low saturation pressure of 2.5 mmHg for krypton at 77 K requires use of a high vacuum apparatus ($\leq 10^{-5}$ Torr). This leads to higher costs for the analysis instrument and to quite sophisticated degas and analysis procedures.

Another approach for routine measurements of surface areas as low as 0.01 m^2/g is to use the Gemini-principle. This patented design effectively compensates for errors arising from the free space in the instrument, temperature gradients, and the changing coolant level. Even more important, the volume adsorbed by the sample is measured directly instead of the calculations performed in the standard design. All other benefits of the static volumetric technique still apply.

Tab. 13-2: Analysis of standard M11-08 with the Gemini

Test No.	Spec. Surface Area (m^2/g)	Test No.	Spec. Surface Area (m^2/g)
1	0,673	6	0,693
2	0,679	7	0,680
3	0,674	8	0,678
4	0,686	9	0,692
5	0,674	10	0,697

The performance of such a system can be checked by analysing a certified reference material, such as No. M11-08 from the Laboratory of the Government Chemist (LGC) in England. The sample is α-alumina powder, with a certified value of (0.6920 ± 0.0145) m^2/g nitrogen B.E.T. surface area. Ten consecutive runs on the Gemini instrument with nitrogen at 77 K resulted in an average of (0.6825 ± 0.0091) m^2/g (see Tab. 13-2), which is in perfect agreement with the certified value.

13.3 Sample Preparation Considerations

Before a sample can be analyzed, it is necessary to remove gas and vapors which may have adsorbed onto the surface from the ambient air. If this is not done, the surface area result can be low and non-reproducible since an indeterminate amount of the surface will be covered with these materials. This step must be done with caution; every effort must be made not to change the original surface of the sample. This sample preconditioning is usually accomplished by either applying a vacuum or purging the sample with an inert flowing gas.

Both methods usually make use of elevated temperatures to hasten the rate at which the contaminants leave the surface. Caution must be used when heating magnesium stearate because melting, dehydration, sintering and decomposition are processes that can drastically alter the surface properties of the sample. A test protocol used to establish that excessive sample preparation temperatures have not been used is to determine the surface area of the sample at successively higher preparation temperatures. Duplication of the results at different temperatures indicates that the initial preparation conditions were satisfactory unless both analyses were performed on a completely degraded sample.

Tab. 13-3: Effect of the degas temperature on the B.E.T. surface area of magnesium stearate

Temperature °C	Spec. Surface Area m^2/g
35	5,138
50	5,106
65	4,834
80	4,542
100	3,651
125	1,581

An example of this sample preparation protocol is seen in Tab. 13-3. One sample was prepared or degassed for four hours at the temperatures indicated. The resulting surface areas are listed to the right of the degassing temperatures. These data are shown in graphical format below the table.

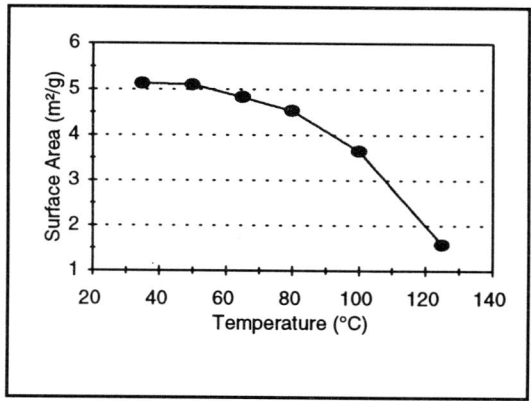

Fig. 13-1: Magnesium stearate B.E.T. surface area as a function of degas temperature

13.4 Analysis of Commercial Magnesium Stearates

Once the sample preparation protocol for magnesium stearate was established, the surface area of four commercially available magnesium stearates was determined. They were degassed for four hours at 35° C, then analyzed using two methods of analysis for comparison. First the more rigorous multipoint analysis was performed and then, for comparison purposes, a single-point analysis of the same samples, under the same testing conditions, was performed. These results are shown in Tab. 13-4 below. The middle column labeled B.E.T. C is indicate of surface energetics.

Tab. 13-4: Magnesium stearate B.E.T. surface area

Sample	Multi-Point-B.E.T. m^2/g	C-Value	1-Point-B.E.T. m^2/g	Diff. rel. %
1	4,923	14,4	4,241	13,9
2	4,286	14,1	3,664	14,5
3	8,056	13,4	6,867	14,8
4	5,957	10,4	5,194	12,8

The multipoint surface area values were calculated from data taken at six relative pressure points (P/P$_0$) equal to 0.05, 0.10, 0.15, 0.25 and 0.30. The single point data were calculated at P/P$_0$ equal to 0.30. The regression value for the multipoint B.E.T. data yielded a correlation coefficient of 0.9998 or greater for all samples. Also, all multipoint analyses were completed in less than 25 minutes.

The difference column in the table above was calculated by subtracting the single-point value from the multipoint value and determining its percentage of the multipoint value. It illustrates the magnitude of the error associated with estimating the surface area by using the single-point technique. This magnitude of error is expected for B.E.T. C values as low as determined here. In general, B.E.T. C values are between 5 and 20 for organic materials. The error in the single-point analysis of these materials is too large for accurate surface data comparison to be made between laboratories or investigators.

13.5 Conclusion

The surface of magnesium stearate by gas adsorption was measured with nitrogen at 77 K. The Gemini-technique allows for automated analysis of these low surface area materials. After setup of a suitable degas procedure to assure optimum sample preparation routine analysis of low surface area materials as magnesium stearate can be performed routinely in a short time.

Address of the author:
Dr. Fritz Metz
micromeritics GmbH
Postfach 10 09 55
Hammfelddamm10
D-41460 Neuss

14 Surface Hydrophobicity - Determination by Rose Bengal (RB) Adsorption Methods

updated reprint from R.H. Müller (1991)

14.1 Importance of Hydrophobicity of Particles and Surfaces for Physical and Biological Interactions

The hydrophobicity of surfaces determines the physical interaction with many compounds, e.g. adsorption of surfactants, polymers and proteins. Hydrophobicity plays also an important role in the interaction of particles or of films and implants with the biological environment. It ranges from the adsorption of body proteins to the attraction of macrophages, e.g. in the process of capsule formation around implants. Biocompatibility of surfaces is also affected by the hydrophobicity, e.g. it determines the chemical nature and quantity of adsorbed body compounds, e.g. opsonins leading to the recognition by macrophages. Quantification of surface hydrophobicity is therefore an important characterisation parameter to judge the performance of systems.

The surface hydrophobicity is one parameter determining the physical surface modification of particulate carriers by polymer adsorption, the interaction of the carriers with cells *in vitro* and also their *in vivo* distribution. The coating of particles by adsorption of polymers depends on the affinity of the polymers to the particle surface (and a possible entropy gain of the system). The polymers are bound by hydrophobic interactions which means that a less hydrophobic particle surface will lead to a reduced affinity of the coating polymers. This results in a thinner or no coating layer at all (Müller, 1991). The surface properties of a thinner film might be different from thick adsorbed layers and subsequently lead to a change in the *in vivo* distribution.

The hydrophobicity of bacteria surface is important for the interaction with phagocytes (van Oss et al., 1975 and 1984). An increase in surface hydrophobicity leads to increased hydrophobic interaction with phagocytes and enhanced phagocytosis. The phagocytosis of yeast particles correlated with their contact angle, that means an increase in surface hydrophobicity led to enhanced adherence and ingestion by the phagocytes, opsonisation increased the surface hydrophobicity and subsequently the ingestion of the yeast particles (Dahlgren and Sunquist, 1981). Albumin particles surface modified with hydrophilic polyethylene glycol showed a reduced uptake in cultures of mouse peritoneal macrophages compared to the more hydrophobic non-modified particles (Artursson et al., 1983). The particles rapidly phagocytosed in the cell cultures exhibited a rapid clearance from the blood in vivo.

The *in vivo* clearance of bacteria from the blood stream is determined by their surface hydrophobicity. Increasing the surface hydrophobicity of *Salmonella typhimurium* 395 MS by coupling of dinitrophenyl ligands to the surface led to an enhanced RES clearance after intravenous injection (Edebo and Richardson, 1985). The hydrophobicity of the bacteria was determined by Hydrophobic Interaction Chromatography (HIC), a technique which was transferred to measure the surface hydrophobicity of drug carriers (Müller, 1991) - as alternative to Rose Bengal adsorption measurements.

The hydrophobicity of the surface determines the interaction with blood components leading to opsonisation and subsequent removal by the RES. Hydrophobic bacteria adsorb aspecifically immunoglobulin G (IgG) and subsequently they are bound to the Fc receptors on phagocytes leading to phagocytosis (van Oss et al., 1984). More hydrophilic bacteria can reduce the unspecific adsorption but will be coated with specific antibodies of the IgG-class (van Oss et al., 1984).

Hydrophilisation of particles by coating with hydrophilic polymers reduces the RES clearance (Müller, 1991). This could be explained by analysing the plasma protein adsorption patterns on particles differing in surface hydrophobicity. In general, hydrophobic surfaces adsorb more total blood protein after i.v. injection than more hydrophilic surfaces. In addition it could be shown that a preferential adsorption of certain plasma proteins occurs (Blunk et al., 1993). The preferentially adsorbed proteins such as Apo E, CII, CIII and AIV are regarded as the responsible factors determining the organ distribution, e.g. affinity to the endothelial cells of the bone marrow by CII. In addition it should be noted that the total amount adsorbed is not necessarily a direct function of hydrophobicity as determined by the RB methods or Hydrophobic Interaction Chromatography. Particles coated with Poloxamers possessing a medium length of ethylene oxide (EO) chains were found to be of medium hydrophobicity compared to a very low hydrophobicity measured for particles coated with Poloxamers possessing long EO chains. However, the total amount of protein was least for the particles of medium hydrophobicity. This was explained by the maximum density of the medium EO chains adsorbed on the surface providing the most effective steric stabilisation barrier (Blunk, 1994). To sum up: hydrophobicity is important but needs to be surely complemented by other characterisation methods such as 2-D PAGE (Blunk, 1994), cf. also chapter 18 about 2-D PAGE.

14.2 The Basic Problem of Measuring Surface Hydrophobicity

A broadly used technique to quantify surface hydrophobicity are contact angle measurements. This is relatively unproblematic for flat surfaces (e.g. films), in case of particles special sample preparation is required. A relationship between the reduction in surface hydrophobicity of coating films as determined by contact angle measurements and RES clearance of i.v. injected polymeric particles was described by Tröster and Kreuter (1988). Contact angle measurements cannot be applied to the hydrated polymer particle in its original dispersion medium. Uncoated particles need to be compressed to form a tablet or the polymer particles need to be dissolved in an organic solvent to cast a polymer film for the contact angle measurement. Alternatively the particle suspension can be placed on a microscope glass, after simple air drying a more or less smooth film is obtained which can be used for the contact angle measurement. The problem is, that relatively hydrophilic particles immediately start to redisperse in the droplet of water used for the contact angle measurement.

Apart from these points the basic problem is the question to which extent the contact angle measurements from a modified sample (e.g. by drying, compressing, dissolution of the polymer and subsequent film casting) have any relation to the true situation of the surface being exposed to a dispersion medium (e.g. hydrated surface in contact to water). In fact, I believe that only artefacts are measured which correlate by chance - just by chance !! - to a certain *in vivo* behaviour. It is similar to the correlation between the decrease in the number of storks in Sweden and the decrease in the birth rate in this country. A really suitable method should allow to quantify surface hydrophobicity of surfaces or surfaces of particulates in their original dispersion medium, at least in conditions with little difference to the original dispersion medium. An approach to this is the adsorption of a hydrophobic dye onto the surface, i.e. measurements of adsorption isotherms and calculation of an affinity constant K or partitioning measurements and calculation of a partitioning coefficient Q. A suitable candidate is Rose Bengal. A special advantage is, that the maximum of adsorption at 542.7 nm changes after adsorption to the surface, that means it allows to measure simultaneously the amount of free and adsorbed Rose Bengal by measuring the absorption of a particle suspension at the two different wavelengths (for details of the basic procedure cf. chapter 17). It is not necessary to remove particles from the dispersion medium by centrifugation when working in very dilute particle suspensions (photometric measurement vs. a control of particle suspension without Rose Bengal). A disadvantage is the high affinity of Rose Bengal to surfaces in general, that means it is necessary to run blanks as control.

Furthermore, coated particles cannot be measured at all by the contact angle method. The surfactants used for the particle coating need to be coated on a film or a plate of the polymeric particle material and to be dried before the measurement (Tröster and Kreuter, 1988). In addition, the basic problem remains, that the properties of the surfactant film adsorbed and dried on this flat polymer film might differ from the properties of a film adsorbed on a strongly curved particle surface. In addition, the surfactant adsorbed on the particle is still hydrated by the surrounding dispersion liquid. Techniques using the adsorption of hydrophobic dyes (e.g. Rose Bengal) appear less suitable for coated particles because of possible displacement of the coating material by the dye. Hydrophobic Interaction Chromatography (HIC) determines the hydrophobicity of the adsorbed surfactant layer on the particle in its original dispersion medium and is therefore regarded to be more suitable for these systems than contact angle and Rose Bengal adsorption measurements (for details cf. chapter 15 to 17).

14.3 Rose Bengal Binding Methods

14.3.1 Rose Bengal Adsorption Isotherms

The binding constant of Rose Bengal to the surface of the standard polystyrene latex particles was used as measure of surface hydrophobicity (Müller et.al, 1986; Mak et al., 1986; Davis et al., 1986). Adsorption isotherms were measured in 0.1 M phosphate buffer (pH 7.4). The latex was incubated with Rose Bengal (3 hours at room temperature), centrifuged for 1 - 2 hours at 20,000 rpm to spin down the particles and the concentration of free Rose Bengal in the supernatant determined spectrophotometrically at 542.7 nm. Due to the high affinity of Rose Bengal for polymer surfaces (e.g. centrifuge tubes) control samples were run in each experiment. The saturated adsorption of Rose Bengal to the latex particles was achieved within the incubation time of three hours. No increased adsorption was found in samples after 24 hours incubation. The binding constant was calculated using a Scatchard plot according to the equation:

$$r/a = KN - Kr$$

r - amount of Rose Bengal adsorbed ($\mu g/\mu m^2$), ($\mu g/mg$)
a - equilibrium concentration of Rose Bengal ($\mu g/ml$)
K - binding constant (ml/μg)
N - maximum amount bound ($\mu g/\mu m^2$), ($\mu g/mg$)

Fig. 14-1 shows the adsorption isotherms obtained using polystyrene latex particles of different sizes (PS-0.06, PS-0.17 and PS-0.90) and surface properties (modified by the introduction of additional amino groups (PS-AR-NH$_2$)). The amount adsorbed per mg latex (dry weight) was plotted against the equilibrium concentration of Rose Bengal for the latex suspensions of different mean particle sizes. The adsorption plateaus are located at very different concentrations of Rose Bengal due to the large difference in the surface area between 0.06 µm and 0.90 µm particles. Plotting the amount of Rose Bengal adsorbed per unit surface area (µg/µm^2) reduced the difference in plateau levels (Fig. 14-2). The difference in the slopes of the adsorption isotherms indicates the different affinity and binding constants of Rose Bengal for the different surfaces. Plotting the data from the slope of the adsorption isotherms using the Scatchard equation (r/a against r) led to straight lines (Fig. 14-3) whereby the slope gives the binding constant K (Tab. 14-1).

Different binding constants were obtained for the small polystyrene latex (0.06 µm) and the larger latices of 0.14 µm and 0.9 µm mean particle size. The small latex were more hydrophobic (K = 0.40 ml/µg) than the larger particles (K = 0.29 and 0.28 ml/µg respectively). This is of importance for the coating of particles with polymers. Despite the fact that the particles of different sizes are made of the same material, they have different surface properties. This is likely to influence their adsorption of polymers, and consequently the thickness and hydrophobicity of the resulting coating layer. The latex with a higher mean particle diameter have a larger number of charged groups on the surface as shown by zeta potential measurements (Müller, 1991). Zeta potential measurements gave a potential of about -65 mV for PS-0.06 but potentials in the range of -75 to -80 mV for the larger latices. The number of charged groups depends on the production conditions used by the manufacturer. The proof that an increased number of surface charged groups reduces the surface hydrophobicity can be seen from the Rose Bengal binding constants obtained for latex after surface chemical modification by the introduction of additional carboxyl groups (PS-COOH) or hydrophilic functional groups (PS-OH, PS-AR-NH$_2$). These latices possess higher zeta potentials and binding constants below 0.28 ml/mg.

Such differences in the surface hydrophobicities of polystyrene latices are important for phagocytosis studies in cell cultures. Cell cultures are being used for the testing of potential coating polymers which are able to protect against *in vivo* phagocytosis (Müller, 1991). Polystyrene latex PS were coated with polymers, and the number of coated particles taken up by the macrophages in the cell culture is compared to an uncoated control. In previous studies (Illum et al., 1987), in order to make detection of the phagocytosed particles easier, 5 µm latex which can be visualised by light microscopy were used. The results obtained in these studies were transferred directly

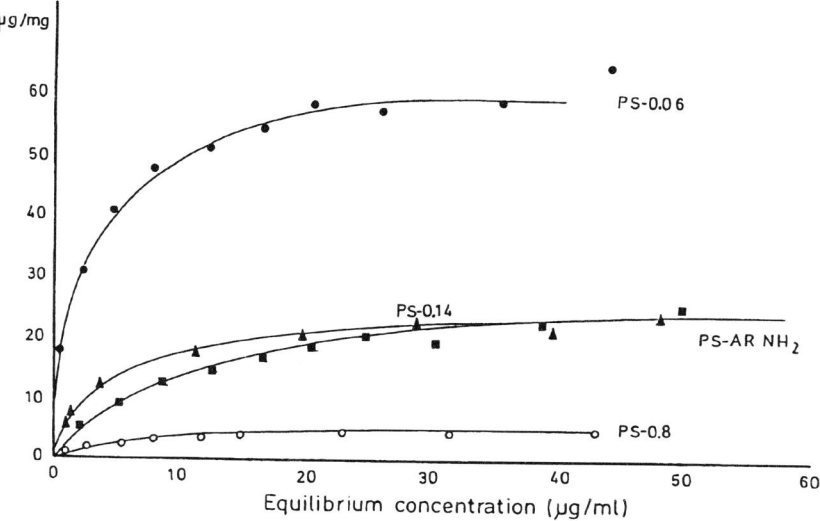

Fig. 14-1: Adsorption isotherm of Rose Bengal on polystyrene particles of 0.06 μm, 0.14 μm, 0.90 μm (PS-0.06, PS-0.14 and PS-0.90) and on polystyrene particles carrying aromatic amino groups on the surface (PS-AR-NH$_2$). The amount of Rose Bengal adsorbed is given per mass polymer (mg).

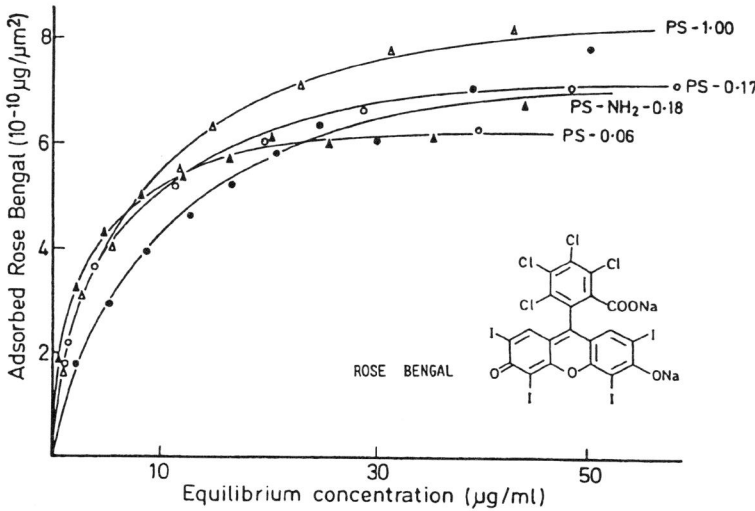

Fig. 14-2: Adsorption isotherm of Rose Bengal on polystyrene particles of 0.06 μm, 0.14 μm, 0.90 μm (PS-0.06, PS-0.14 and PS-0.90) and on polystyrene particles carrying aromatic amino groups on the surface (PS-AR-NH$_2$). The amount of Rose Bengal adsorbed is given per unit of surface area (μm^2).

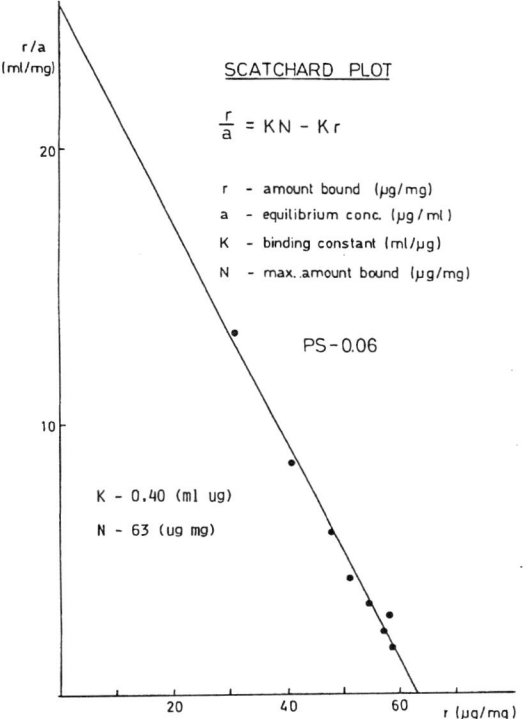

Fig. 14-3: Scatchard plot of the adsorption isotherm of Rose Bengal on 0.06 µm polystyrene particles (PS-0.06).

Tab. 14-1: Binding constants K, maximum amount bound N per mg latex and per unit surface area for the standard latex particles obtained from Rose Bengal adsorption isotherms (PS- polystyrene, PSF - polystyrene fluorescently labelled; AL-NH_2 (aliphatic amino groups), AR-NH_2 (aromatic amino groups), COOH, OH - functional groups introduced to the particle surface, figure behind abbreviation: size in µm).

Latex particles	binding constant K (ml/mg)	maximum amount N bound per	
		mass polymer (µg/mg)	unit surface (10^{10} µg/µm^2)
PS-0.06	0.40	63.0	6.6
PS-0.14	0.29	24.1	5.9
PS-0.90	0.28	4.5	7.0
PSF-0.19	0.12	22.3	5.5
PS-AL-NH_2-0.19	0.23	20.9	6.9
PS-AR-NH_2-0.18	0.14	24.3	7.6
PSF-COOH-0.21	0.27	22.5	6.3
PS-COOH-0.19	0.06	3.1	1.0
PS-OH-0.25	0.04	4.1	1.8

to 60 nm latex. Because of the difference in the surface hydrophobicities of the latex preparations, the properties of the coating layers are likely to differ. The extrapolation of results obtained with 5 µm particles to 60 nm particles is therefore not necessarily valid.

Fluorescent polystyrene particles are produced by attachment of hydrophobic fluorescent dyes to the naked particle surface. For such fluorescently labelled particles it was expected that the Rose Bengal binding constant would be around 0.30. Such a high binding constant could be found for the fluorescent labelled carboxylated particles PSF-COOH. The attachment of the hydrophobic dye yellowish green increased the binding constant from $K = 0.06$ for the unlabelled PS-COOH to $K = 0.27$ for PSF-COOH (batch A). However, a purchased second batch of PSF-COOH (batch B) was much less hydrophobic than the first batch indicated by a low binding constant ($K = 0.13$ ml/mg). From the Rose Bengal measurements this result could not be explained. Only investigation of the two batches by Hydrophobic Interaction Chromatography (cf. chapter about HIC) could explain the observed differences in surface hydrophobicity.

The introduction of functional groups to the particle surface led to a reduction in the surface hydrophobicity and the measured binding constant K. Attachment of aromatic and aliphatic amino groups on the surface reduced K to 0.23 and 0.14 respectively, while introduction of carboxyl and hydroxyl groups reduced K even further, to 0.06 and 0.04 ml/µg respectively. It is thought, that the small reduction in K for the particles after attachment of amino groups might be due to the fact that the groups are attached via an aliphatic chain or an aromatic ring. This might bring about an increase in the surface hydrophobicity which compensates the decrease produced by the attached functional groups. An explanation as to why the particles with the aliphatic amino groups were more hydrophobic than the ones with the aromatic amino groups could not be given because of the lack of information available from the supplier Polysciences about the chemical modification process.

The objective of the characterisation of the latex was to find suitable methods for the detection of differences in surface hydrophobicities. Knowledge of the reasons for the observed differences in hydrophobicity was not essential for the development of a suitable measuring technique. For further studies on the differences in polymer coatings, it was essential to have a detailed knowledge of the production method and the chemicals involved. Contamination of the particles by adsorbed detergents is important as this may effect coating by polymers, e.g. Poloxamer and Poloxamine. It was shown that commercial latices contained surfactants despite the fact that they were sold as surfactant free (Müller, 1991).

14.3.2 Rose Bengal Partitioning Method

The determination of adsorption isotherms is time consuming, so a faster partitioning method was investigated (Müller et. al, 1986). The latex suspension is regarded as a two phase system, whereby the surface of the particles is one phase and the dispersion medium is the second. Rose Bengal partitions between these two phases and a partition quotient PQ can be determined:

$$PQ = \frac{\text{amount Rose Bengal bound on surface}}{\text{amount Rose Bengal in dispersion medium}}$$

For the partitioning experiments, the concentration of Rose Bengal was kept constant and the concentration of latex particles was varied. Increasing the latex concentration leads to an increase in the surface area and consequently the partition quotient should increase linearly. Plotting PQ against the particle surface area will therefore yield a straight line from which the slope S can be taken as a measure of hydrophobicity.

For a partitioning analysis, suspensions with constant Rose Bengal concentration (20 µg/ml) but increasing latex particle concentrations were prepared. The suspensions were incubated for three hours, the particles centrifuged at 20,000 rpm and the amount of free Rose Bengal in the supernatant was determined spectrophotometrically at 542.7 nm.

Plotting of PQ against the total surface area of the latex gave a line with a very steep slope for the most hydrophobic polystyrene latex (PS-0.06, PS-0.14 and PS-0.9), when compared to the relatively hydrophilic carboxylated latex PS-COOH (Fig. 14-4). A good correlation was found between the binding constants K, obtained from the Rose Bengal adsorption isotherms, and the slopes S obtained by the partitioning method. Plotting S versus K produced a straight line. The particles are placed in order of increasing hydrophobicity (Fig. 14-5). The distinct difference in surface hydrophobicity between PS-0.06 and the larger polystyrene latices (0.14 and 0.90) could be confirmed. Tab. 14-2 gives an overview of the data obtained using both Rose Bengal methods.

Hydroxylated and carboxylated latex are at the lower edge of the surface hydrophobicity range. The affinity of Rose Bengal for their surfaces is very low. Consequently, high concentrations of Rose Bengal had to be used for the determination of their adsorption isotherms and large quantities of particles were required for the partitioning experiment in order to obtain a detectable amount of Rose Bengal adsorbed onto the particle surface. This shows the limitations in the application of the method to particles which are more hydrophilic than PS-OH and PS-COOH.

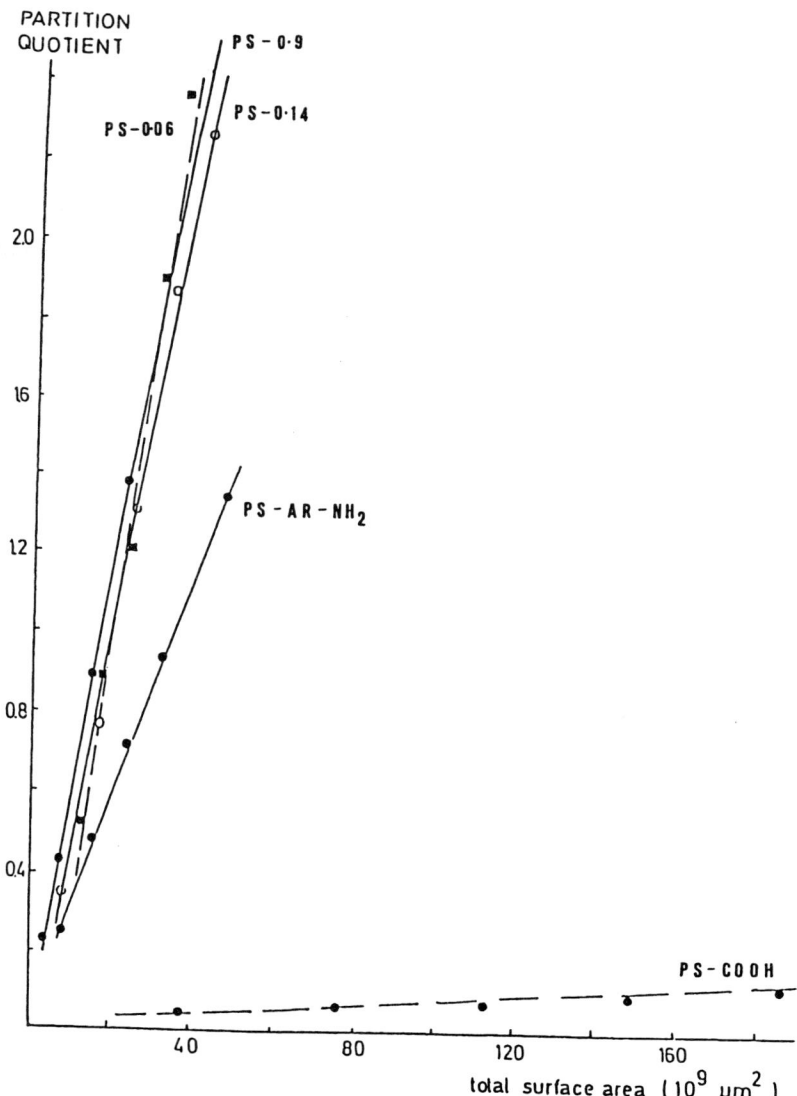

Fig. 14-4: Rose Bengal partitioning method. Plot of the partition quotient versus the total surface area of the particles. The slope S of the straight line is a measure of the surface hydrophobicity of the latex (Müller et al., 1986).

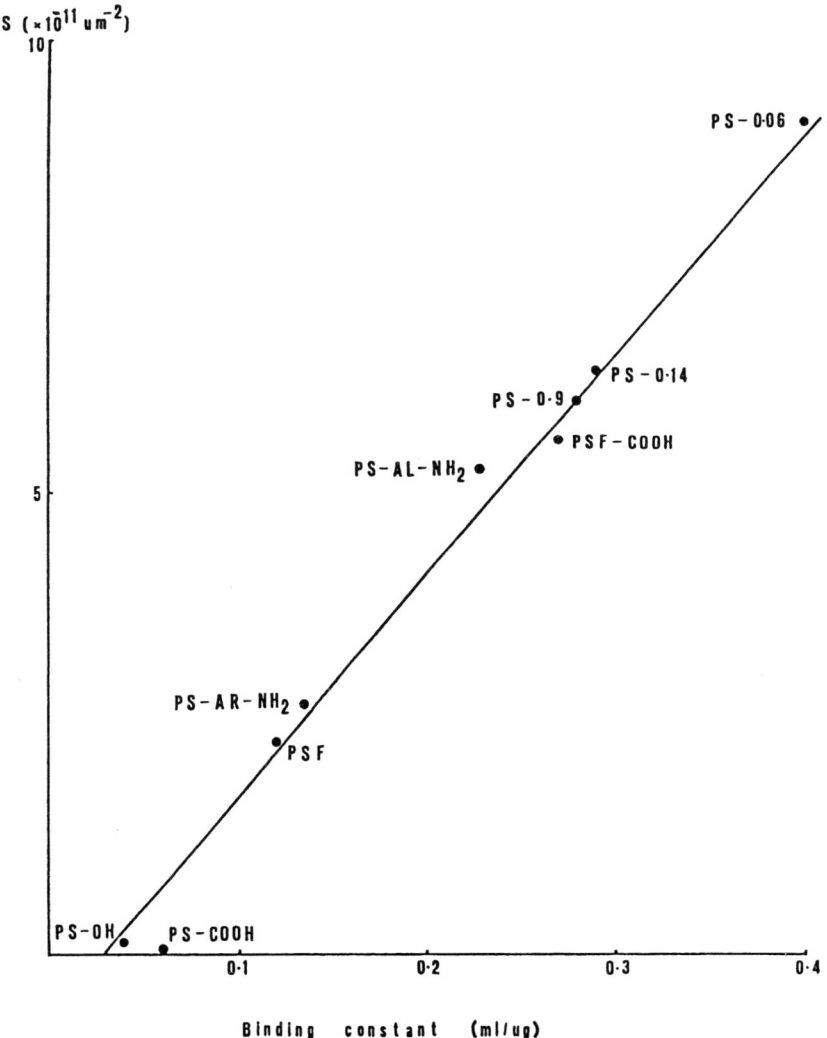

Fig. 14-5: Plot of the slope S (from the partitioning method) versus the binding constant K (from the adsorption isotherms) for the series of standard latex. The latex are placed in order of increasing surface hydrophobicity (Müller et al., 1986).

Tab. 14-2: Characterisation data from both Rose Bengal methods (after Müller et al., 1986; Davis et al., 1986).

Particles (figure = size in µm)	k (ml/µg)	N (µg/mg)	total surface area (µm^2/ng) x E10	max. amount bound/unit surf. area (µg/µm^2) x E10	slope from part. exp. (ml/µm^2) S
PS-0.06	0.40	63.0	9.52	6.61	9.124 E-11
PS-0.14	0.29	24.1	3.36	7.18	6.39 E-11
PS-0.9	0.28	4.5	0.63	7.02	6.023 E-11
PSF-0.19	0.12	22.3	3.01	7.43	2.335 E-11
PS-AR-NH$_2$-0.18	0.14	24.3	3.21	7.58	2.723 E-11
PS-NH$_2$-0.19	0.23	20.9	3.01	6.94	5.29 E-11
PSF-COOH-0.21	0.27	22.5	2.72	8.27	5.56 E-11
PS-COOH-0.19	0.06	3.1	3.01	1.02	5.131 E-13
PS-OH-0.25	0.04	4.1	2.29	1.82	1.42 E-12

14.4 Summary

Rose Bengal is a suitable probe to measure the surface hydrophobicity of uncoated model carriers and particles in general. Determination of the binding constant of a range of particles (via the production of adsorption isotherms) showed an increased adsorption with increasing particle hydrophobicity. The Rose Bengal binding constants were used to place the particles in order of increasing surface hydrophobicity. The same result could be obtained using a quicker dye-partitioning method.

Differences in the surface hydrophobicity of particles made from the same material (polystyrene) but differing in size (over the range 60 nm to 0.9 µm) were found. This questions the validity of interpreting results obtained with larger particles (in surface coating and phagocytosis studies) in respect of small latex particles.

Large differences between batches of latex were found. This was especially true for surface modified particles (fluorescent label on surface). Obviously differences in the chemical reaction for the surface modification led to particles very different in hydrophobicity. The variations in hydrophobicity between particles of different sizes or between batches show the importance of characterisation of model latices. Otherwise observed effects in an experiment due to batch differences in surface properties might lead to wrong conclusions.

A limitation of the Rose Bengal methods is that the calculated binding constant K or the slope S are only average measures of the hydrophobicity of the particles. They do not give any information about possible subpopulations, which may differ in surface hydrophobicity. Furthermore, the methods are not applicable to particles less hydrophobic than PS-COOH or PS-OH and to particles coated with hydrophilic polymers (e.g. Poloxamer, possible displacement of polymer).

14.5 References

Artursson, P., Laakso, T. and Edman, P., Acrylic microspheres *in vivo*: Blood elimination kinetics and organ distribution of microparticles with different surface characteristics, J. Pharm. Sci. 72, 1415-1420 (1983)

Blunk, T., Hochstrasser, D.F., Sanchez, J.-C., Müller, B.W. und Müller, R. H., Colloidal carriers for intravenous drug targeting: Plasma protein adsorption patterns on surface-modified latex particles evaluated by two-dimensional polyacrylamide gel electrophoresis, Electrophoresis 14, 1382-1387 (1993)

Blunk, T., Plasmaproteinadsorption auf kolloidalen Arzneistoffträgern - Analytik, Korelation mit Oberflächeneigenschaften, Implikationen für das Drug Targeting, PhD thesis, University of Kiel, Germany, 1994

Dahlgren, C. and Sunquist, T., Phagocytosis and hydrophobicity: A method of calculating contact angles based on the diameter of sessile drops, J. Immunol. Meth. 40, 171-179 (1981)

Davis, S.S., Douglas, S., Illum, L., Jones, P.D.E., Mak, E. and Müller, R.H., Targeting of colloidal carriers and the role of surface properties, in Targeting of Drugs with Synthetic Systems (Gregoriadis, G., Senior, J. and Poste, G., eds.), Plenum Press New York, 123-146 (1986)

Edebo, L. and Richardson, N., Enhancement of hydrophobic interaction, negative charge and phagocytosis by dinitrophenyl ligand coupling to *Salmonella typhimirium* 395 MS, Int. Archs. Allergy appl. Immun. 78, 345-352 (1985)

Illum, L., Jacobsen, L.O., Müller, R.H., Mak, E. and Davis, S.S., Surface characteristics and the interaction of colloidal particles with mouse peritoneal macrophages, Biomaterials 8, 113-117 (1987)

Mak, E., Davis, S.S., Illum, L. and Müller, R.H., Determination of surface properties of 'standard' latex particles, Abstr. British Pharmaceutical Conference, 22-24 September Jersey, 100P (1986)

Müller, R.H., Davis, S.S., Illum, L. and Mak, E., Particle charge and surface hydrophobicity of colloidal drug carriers, in Targeting of Drugs with Synthetic Systems (Gregoriadis, G., Senior, J. and Poste, G., eds.) Plenum New York, 239-263 (1986)

Müller, R.H., Colloidal Carriers for Controlled Drug Delivery and Targeting - Modification, Characterisation and In Vivo Distribution, Wissenschaftliche Verlagsgesellschaft Stuttgart and CRC Press Boca Raton , 1991

Tröster, S.D. and Kreuter, J., Contact angle of surfactants with a potential to alter the body distribution of colloidal drug carriers on poly(methyl methacrylate) surfaces, Int. J. Pharm. 45, 91-100 (1988)

Van Oss, C.J., Gillman, C.F. and Neumann, A.W., Phagocytic engulfment and cell adhesiveness as cellular surface phenomena, Marcel Dekker New York (1975)

Van Oss, C.J., Absolom, D.R. and Neumann, H.W., Interaction of phagocytes with other blood cells and with pathogenic and nonpathogenic microbes, Ann. N.Y. Acad. Sci. 416, 332-350 (1984)

Address of the author:
Prof. Dr. Rainer H. Müller
Department of Pharmaceutics, Biopharmaceutics & Biotechnology
Free University of Berlin
Kelchstr. 31
D-12169 Berlin

15 Theory and Set Up of Hydrophobic Interaction Chromatography (HIC)

Dr. K. Thode and Prof. Dr. R. H. Müller, Free University of Berlin

15.1 Theory of HIC

HIC is a liquid chromatography technique that has expanded rapidly since its development in the early 1970`s. It was primarily used for purifying proteins but was also applied to other biotechnological areas, e.g. receptor studies [1], the determination of surface hydrophobicity of bacteria [2], bacteria toxins [3], and blood cells [4]. It was also transferred to model drug carriers [5, 6].

A polymer matrix is a fundamental requisite for this technique. Mostly, alkyl-agarose is used as matrix in the stationary phase. Immobilized ligands determining the degree of hydrophobicity are attached to the agarose. Regarding alkyl chains as commonly used ligands an increase in chain length leads to an increase in hydrophobicity. Many possibilities exist for using different ligands and matrices, the mechanism of coupling may also be varied [7]. It is obvious that the stationary phase covers a wide range of hydrophobicity. The most important factors concerning the interaction of samples with the hydrophobic ligands are the sample itself, the hydrophobicity of the adsorbent, its degree of substitution and the composition of the solvent and elution medium, respectively. Hjertén [8] formulated a widely accepted theory that is based on the second law of thermodynamics:

$$\Delta G = \Delta H - T\Delta S \qquad \text{Eq. 15-1}$$

In aqueous dispersions hydration of molecules prevents hydrophobic interaction. An increased salt concentration changes the arrangement of water molecules. Consequently, ordered water molecules surrounding the proteins as well as the ligands are removed and rearrange around the ions added to the solvent/elution medium. This leads to partially hydrated ligands and proteins, hydrophobic areas may interact. Following up the second law of thermodynamics, entropy (ΔS) increases because of the displaced water molecules. The resulting negative value for the change in free energy (ΔG) of this system causes a thermodynamically favourable interaction between proteins and the hydrophobic ligands [7].

On the other hand hydrophobic interaction is also influenced by other factors:

a. **Temperature**: An increase in temperature leads to an increase in hydrophobic interaction, in agreement with the second law of thermodynamics. This was proved by experiments carried out e.g. by Jennissen [9]. Performing HIC should therefore of course take place at a constant temperature.

b. **PH-value**: It could be observed that a decrease in pH-value results in an increase in hydrophobic interactions and vice versa [10]. This is due to the fact that hydrophilicity of the proteins and titration of charged groups are influenced correspondingly. Nevertheless, a stable retention of proteins is assessed in the range of pH 5 - 8.5 [11]. Using a medium with a buffered pH-value of 6.8 minimizes the dissociation of carboxy- and sulfate-groups which are present on the gel matrix in a very small quantity [12].

c. **Type and concentration of salt**: The presence of salts in the elution buffer is a driving force for hydrophobic interaction. Ionic strength exceeding 0.02 is required to avoid hydrophilic and to promote hydrophobic interaction [12]. Despite this, the type of salt is important, too. Salts showing a chaotropic effect diminish hydrophobic interaction, whereas salts showing a salting-out effect do the opposite. This is caused by the changes in the order of water molecules as described above. The effects of anions and cations on the order of water molecules are described in the Hofmeister series. An overview is given by Melander et al. [13].

d. **Presence of additives**: Water miscible alcohols and detergents, for example, may falsify the HIC results. Competing effectively with bound or not yet bound proteins these molecules may occupy immobilized ligands resulting in an increase of rapidly eluted proteins. Therefore, one has to make sure, that dispersions of particles coated with surfactants do not contain any excess of surfactant. Such additives as well as less concentrated salt solutions are used to purify HIC columns and to determine the fraction of the sample bound to the matrix.

The factors influencing HIC are shown in Fig. 15-1.

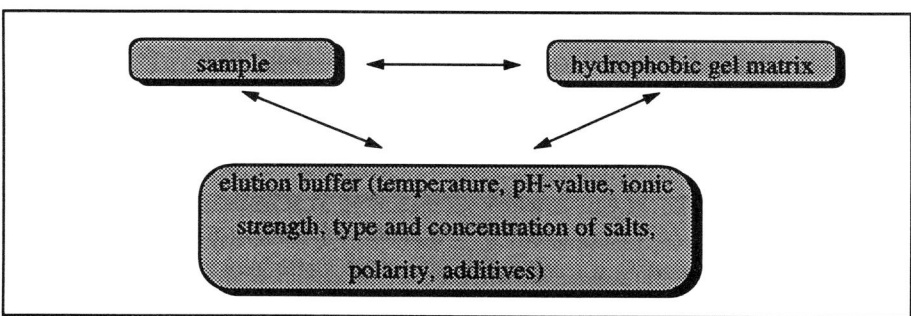

Fig. 15-1: Interactions between the gel medium, the sample to be analyzed and the elution buffer

15.2 Running HIC

Before starting HIC the hydrophobic gel medium, the elution buffer and the washing medium should be selected. For example, phosphate buffer pH 6.8 is a widely used elution medium. It contents 1.78 g Na_2HPO_4 x $2H_2O$, 1.56 g NaH_2PO_4 x $2H_2O$ and 7.36 g NaCl dissolved in one litre of double distilled water. Furthermore, the wavelength for the detection of the sample has to be determined. Disturbances caused by additives have to be excluded. A suitable sample volume is received when focusing on the absorption. The chosen volume should guarantee a maximum absorption of 1.0, complete elution in one fraction presumed.

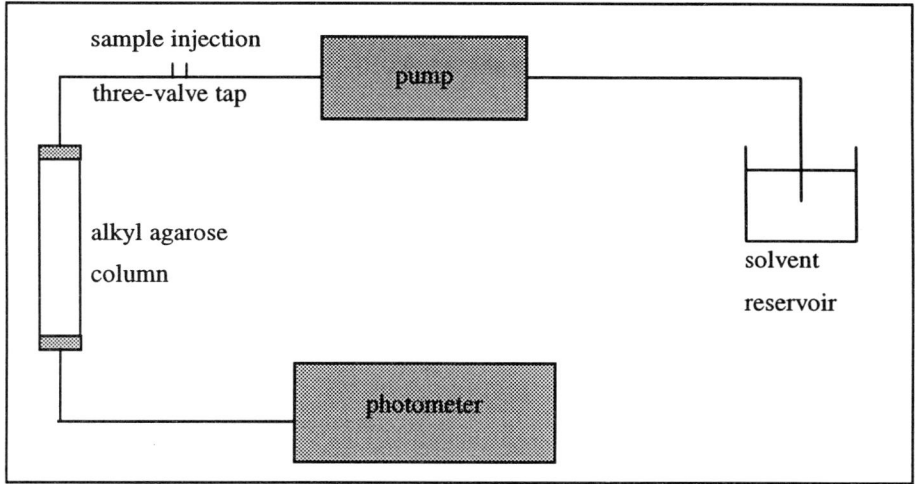

Fig. 15-2: Experimental design of hydrophobic interaction chromatography (HIC)

In Fig. 15-2 the set up of HIC is shown. The system consists of a pump, a glass column, wherein the matrix is packed to a column height of 5-15 cm (the diameter of the column is 1 cm), and a UV/VIS-spectrometer with the option of on-line work. A suitable speed for pumping the elution buffer into the system is about 0.3 ml/min. After equilibration of the column using double column volume of buffer, HIC starts when the sample is applied by opening the three-valve tap. Unbound particles are eluted with the void volume of the column. Particles interacting with the column are eluted retarded, i.e. elution volume > void volume. Particles bound to the column may be washed off by using a washing solution, e.g. 0.1% Triton X-100 in buffer. Displacement will not be successful in case of very strong interactions between the sample and the gel medium. The column is cleaned by flushing with double distilled water, alcohol-water mixtures can also be employed (cf. chapter 16).

There are three possible subpopulations which can be expected: Particles passing the matrix without interaction or being retarded, determined as the area under the elution peak. Particles retained in the column and being eluted by washing solution, determined as the area under the wash peak. Finally, particles not eluted neither by buffer nor by washing medium. This fraction can be calculated from the theoretical area for 100% elution subtracting the AUC of the elution peak with buffer and the AUC of the peak when washing with Triton X-100. Beside the relationship between these areas, the volume required until the elution of unbound particles starts (retention volume) can be used to quantify surface hydrophobicity.

15.3 Running Mini - HIC

A modified version of HIC was developed by Wallis and Müller [14]. It could be applied also to magnetite dispersions (cf. chapter 17) [15]. Having its advantages in reduced time and costs, it is suitable for screening the surface hydrophobicity of samples. Figure 3 shows the set up of Mini-HIC.

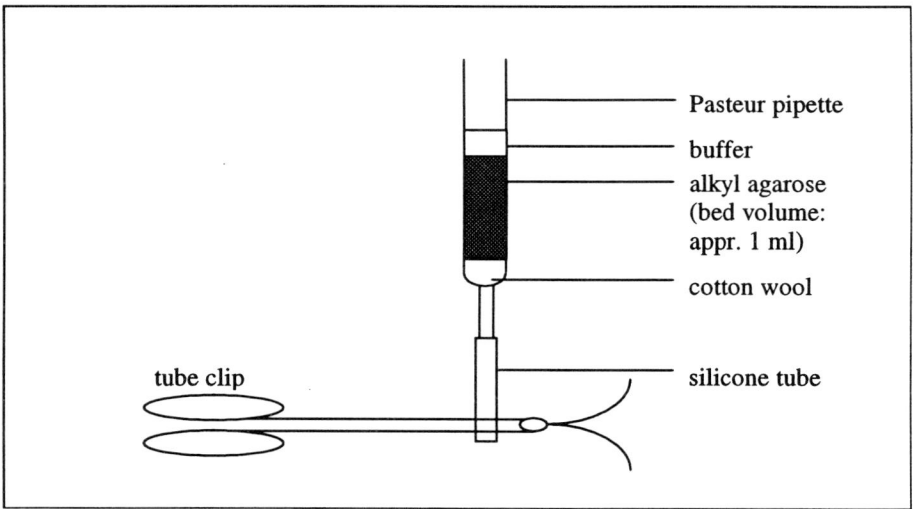

Fig. 15-3: Experimental design of mini - hydrophobic interaction chromatography (Mini-HIC)

For carrying out Mini-HIC a silicone tube is attached to the tip of a Pasteur pipette, the tube is locked by a tube clip. After having filled the outlet of the pipette with a tiny piece of cotton wool 1.0 ml of the hydrophobic gel medium is filled into this system and equilibrated. The sample is directly layered onto the matrix, immediately followed by elution. In contrast to HIC, there is no on-line work to detect fractions of bound and unbound particles. The fractions are collected in distinct fractions (for example 0.33 ml) and detected using a UV/VIS-spectrometer. Due to the relatively small bed volume, the different retention volumes cannot be resolved with high resolution as on large scale HIC. However, when dividing the eluted volume into fractions of 333 mg ($^1/_3$ ml) collected in small-volume cuvettes, a sufficiently high resolution for screening purposes can be achieved. To control the eluted volume, the cuvette can be placed on a digital balance. Distinct advantages of Mini-HIC are:

- extremely short time for analysis allowing screening of large sample numbers
- no possible artefacts due to particles remaining on the column from previous analysis (not removable by Triton X-100 from column) because after one analysis the columns will be discarded
- cost-effective

For obtaining more detailed information about the surface hydrophobicity of a particle population, large column HIC is recommended. Suitable column volumes are 10 ml. Especially subpopulations in hydrophobicity can be analyzed with high resolution (cf. chapter 16).

15.4 References

[1] Kuehn, L., Meyer, H., Reinauer, H., Hydrophobic interaction chromatography as a tool in insulin receptor study, Proc. 2nd Int. Insulin Symp. (1980) 243 - 250

[2] Magnusson, K.-E., Stendahl, O., Stjernström, I., Edebo, L., The effect of colostrum and colostral antibody SIgA on the physicochemical properties and phagocytosis of *Escherichia coli* O86, Act. Pat. S. Sect. B86 (1978) 113 - 120

[3] Schoel, B., Welzel, M., Kaufmann, S.H.E., Hydrophobic interaction chromatography for the purification of cytolytic bacterial toxins, Journal of Chromatography A, 667 (1994) 131-139

[4] Matsumoto, U., Shibusawa, Y., Surface affinity chromatographic separation of blood cells. V. Retention behaviour of human peripheral blood cells on poly(propylene glycol) bonded agarose columns and its relationship to surface hydrophobicities of the gel beads and the cells, Journal of Chromatography, 356 (1986) 27 - 36

[5] Blunk, T., Mak, E., Müller, R.H., Characterization of colloidal drug carriers: Determination of surface hydrophobicity by hydrophobic interaction chromatography, Pharm. Ind. 55 (6) (1993) 612 - 615

[6] Müller, R.H. (Ed.), Colloidal carriers for controlled drug delivery and targeting, modification, characterization and in vivo distribution, Wissenschaftliche Verlagsgesellschaft, Stuttgart (1991) 110 - 123

[7] Pharmacia, Hydrophobic interaction chromatography - principles and methods, Pharmacia, Uppsala, Sweden (1993)

[8] Hjertén, S., Fractionation of proteins by hydrophobic interaction chromatography, with reference to serum proteins, Proceedings Int. Workshop on Technology for Protein Separation & Improvement of Blood Plasma Fractionation, Reston, Virginia (1977) 410 - 421

[9] Jennissen, H.P., Multivalent interaction chromatography as exemplified by the adsorption and desorption of skeletal muscle enzymes on hydrophobic alkyl-ligands, Journal of Chromatography 159 (1978) 71 - 83

[10] Hjertén, S., Some general aspects of hydrophobic interaction chromatography, Journal of Chromatography 403 (1987) 85 - 98

[11] Hjertén, S., Yao, K., Eriksson, K.-O., Johansson, B., Gradient and isocratic high performance hydrophobic interaction chromatography of proteins on agarose columns, Journal of Chromatography 359 (1986) 99 - 109

[12] Pharmacia, Gel filtration - theory and practice, Pharmacia Fine Chemicals, Freiburg (1984)

[13] Melander, W., Horvath, C., Salt effects on hydrophobic interactions in precipitation and chromatography of proteins: an interpretation of the lyotropic series, Arch. Biochem. Biophys. 183 (1977) 200 - 215

[14] Wallis, K.H., Müller, R.H., Determination of the surface hydrophobicity of colloidal dispersions by Mini - hydrophobic interaction chromatography, Pharm. Ind. 55, 12 (1993) 1124 - 1128

[15] Thode, K., Kresse, M., Pfefferer, D., Semmler, W., Müller, R.H., Mini - HIC for the determination of the surface hydrophobicity of magnetite dispersions, Proc. 1st World Meeting APGI/APV, Budapest (1995) 509 - 510

Address of the authors:
Dr. Kai Thode and Prof. Dr. Rainer H. Müller
Department of Pharmaceutics, Biopharmaceutics & Biotechnology
Free University of Berlin
Kelchstr. 31
D-12169 Berlin

16 Hydrophobic Interaction Chromatography (HIC) for Determination of the Surface Hydrophobicity of Particulates

updated reprint from R.H. Müller (1991)

16.1 General Applications of HIC

Hydrophobic Interaction Chromatography (HIC) is a column chromatography which separates substances or particulates on the basis of differences in their hydrophobic interaction with a hydrophobic gel matrix. To avoid an overlapping of ion-exchange, charge and hydrophobic effects the matrices need to be composed of a neutral gel, such as alkyl-sepharose CL-4B. The separation achieved depends on the hydrophobicity of the solute, the hydrophobicity of the matrix and the interactions with and between the solvent water molecules. Any perturbation which effects one of these components (e.g. ionic strength of the water and temperature) changes the separation achieved, providing a great degree of flexibility in the design of the experimental conditions. Addition of salts can promote or reduce the hydrophobic interactions between matrix and solute. Addition of anions such as Cl^- and even more PO_4^{3-} increase hydrophobic interactions ("salting out" effect), whereas the addition of cations such as Ca^{2+} and Ba^{2+} disrupts the structure of water ("chaotropic" effect) and leads to a relative decrease in interactions (von Hippel and Schleich, 1969; Pahlmann et al., 1977). Solutes can be bound to the matrix in the presence of a high ionic strength (e.g. 4 M NaCl) and eluted by lowering the salt concentration or changing to an ion with a lower salting out effect. Alternatively, non-ionic detergents such as Triton X-100 can be used. The hydrophobic region of the detergent molecule binds to the matrix and displaces the solutes or particulates bound to the matrix.

16.2 Experimental

The apparatus used for HIC consisted of a column (bed volume 10 ml), an arrangement of two solvent pumps and a UV spectrometer connected to a chart recorder for detection of the particles (e.g. at 400 nm) (Fig. 16-1). To create a surfactant gradient, 0.1 % Triton X-100 was pumped from the reservoir (R) to the mixing container (M) at a flow rate of 0.3 ml/min. Before the start of the experiment, the column was

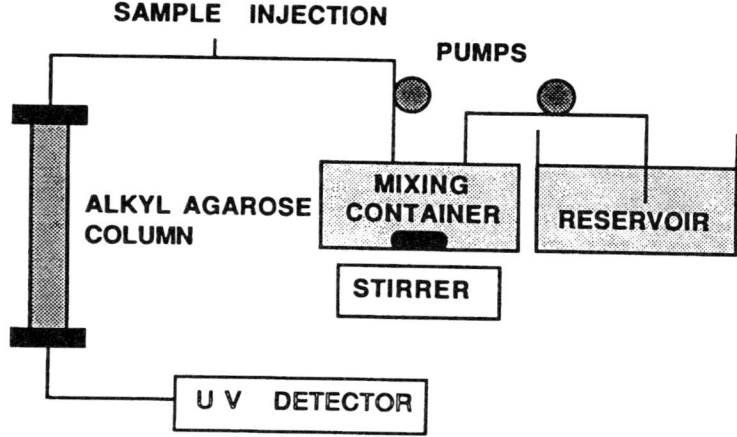

Fig. 16-1: Hydrophobic Interaction Chromatography apparatus. In order to create a Triton X-100 gradient, 0.1 % surfactant solution was pumped from the reservoir (R) to the mixing container (M).

equilibrated with buffer (0.02 M phosphate buffer in 0.3 M NaCl). Latex particles (15 µl of a 2 % w/v suspension) were loaded onto the column and eluted at a flow rate of 0.3 ml/min. Washing of the column was performed using a sequence of different solutions: 10 ml of 0.1 % Triton X-100 in water, then 10 ml of distilled water, 10 ml of 10 % ethanol, 20 ml of 5 % butanol and finally equilibration with 10 ml of distilled water (flow rate 0.3 ml/min).

Gel matrices with increasing hydrophobicity (increasing length of the coupled alkyl chain) have been employed:

>Sepharose CL-4B (neutral agarose)
>Agethane (ethyl-agarose)
>propyl-agarose (propylamine-agarose)
>butyl-agarose (butylamine-agarose)
>pentyl-agarose (pentylamine-agarose)
>hexyl-agarose (hexylamine-agarose)
>octyl-agarose (octylamine-agarose)

The model carriers (polystyrene latex particles) were loaded onto the column in 0.3 M NaCl adjusted to pH 6.8. This pH minimizes charge effects of the column. Higher salt concentrations led to flocculation of the uncoated particles on the column as shown by PCS determinations of the eluted particles. In general it is recommended to measure the size of eluted particulates. This allows to exclude flocculation of particles or coalescence of emulsions and liposomes. In case of particles surface-modified by adsorption of polymers (e.g. poloxamer and poloxamine series) possible desorption of polymer can be detected. A Triton X-100 gradient proved to be suitable for the elution of uncoated model drug carriers (polystyrene latex particles). Latex particles surface-modified by adsorption of polymers showed a weaker interaction with the matrix and were eluted with the buffered saline. The elution volume increased with increasing hydrophobicity of the coating layer. To differentiate between coatings of similar low hydrophobicity, the hydrophobicity of the matrix was increased, i.e. alkyl-agaroses with increased length of the alkyl chain were used.

16.3 Characterization of Polystyrene Latex Particles with Different Functional Surface Groups

To investigate the suitability of HIC for the characterization of unmodified (i.e. non-coated) drug carriers, a series of polystyrene (PS) standard latices were used as model particles (Tab. 16-1 (Müller, et al., 1986)). The particles were loaded onto the column in the presence of buffer. The carboxylated (PS-COOH) and the hydroxylated particles (PS-OH) could be eluted (Fig. 16-2) without using Triton X-100. Both passed down the column without any interaction (elution volume of appr. 4 ml which

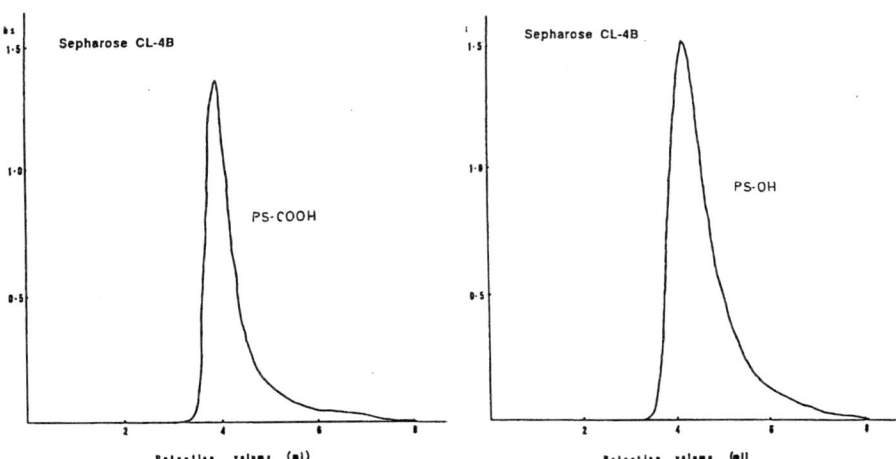

Fig. 16-2: Chromatogram of carboxylated (PS-COOH) and hydroxylated (PS-OH) polystyrene latex particles on Sepharose CL-4B. The latex pass down the column without interaction (elution volume = void volume).

is equal to the void volume of the column). The other particles of the series were bound to the gel matrix and could only be eluted by using 0.1 % Triton X-100. This indicates that the hydroxylated and carboxylated latex had less hydrophobic surfaces.

To be able to detect differences between the surface hydrophobicity of PS-OH and PS-COOH, a more hydrophobic matrix (ethyl-agarose) was used. The more hydrophobic ethyl-matrix was able to bind these particles due to increased hydrophobic interaction. The elution of the particles was performed using a Triton X-100 gradient (0 % - 0.025 %).

The chromatogram obtained from polystyrene latex (PS-0.14) showed a broad peak from an elution volume of 27 - 40 ml (Fig. 16-3, upper curve) indicating that there are differences in surface properties within the particle population. The hydroxylated and the carboxylated particles gave a main peak at 20.0 ml and 22.6 ml respectively (Fig. 16-3). This indicates only a slight difference in surface hydrophobicity as detectable by HIC. There is however a second peak at 32.0 ml for PS-OH and a long, distinct tail for PS-COOH. That proves that within the particle population there is a relatively hydrophilic fraction and a second fraction similar in surface hydrophobicity to the unmodified polystyrene particles (PS-0.14). These results can be easily explained by looking at the production method of these particles, as employed by Polysciences. The polystyrene latices are firstly polymerized and then the surface is modified by the chemical introduction of additional hydroxyl or carboxyl groups. This reaction will of course not lead to surface modification of all particles to the same extent. The second peak at around 32.0 ml can therefore be attributed to polystyrene latex particles with little or no surface modification.

The efficiency of the chemical surface modification seemed to vary from batch to batch, as in different batches a varying ratio of the main 20 ml peak to the minor peak (at around 32 ml) was obtained. An extremely large batch to batch variation was found during the investigation of two batches of fluorescent-labelled carboxylated latex (PSF-COOH) (Mak et al., 1986). The chromatograms also exhibited two peaks; the first peak can be attributed to a hydrophilic fraction and the second to a much more hydrophobic fraction. Batch A shows the main peak at 31.4 ml and a minor peak at 16.3 ml (Fig. 16-4). However, the main peak of batch B is located at 16.3 ml indicating that batch B is less hydrophobic than A. The coupling reaction of the hydrophobic dye yellowish green with the latex was more successful in batch A than in batch B. In the Rose Bengal adsorption studies, a low binding constant was obtained for the more hydrophilic batch B. It was expected to find a higher binding constant due to the attachment of the hydrophobic dye on the particle surface. Such expected high binding constant could only be found for batch A. These differences in the binding constants and related surface hydrophobicity of the particles found from the Rose

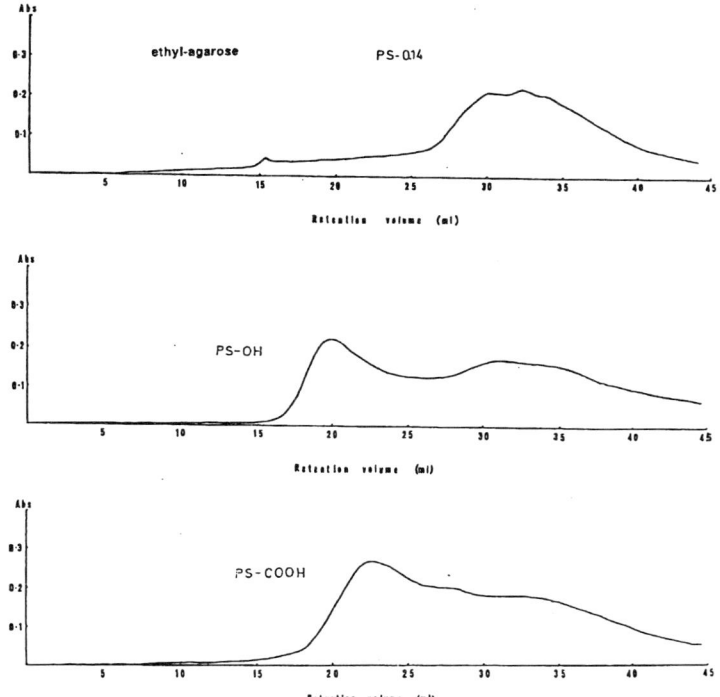

Fig. 16-3: Chromatograms of latex on ethyl-agarose gel: polystyrene (PS-0.14), hydroxylated polystyrene (PS-OH) and carboxylated polystyrene (PS-COOH) particles. Elution was performed using a Triton X-100 gradient.

Bengal studies could not be explained when using this characterization method. A binding constant, as determined by the Rose Bengal method, is only an average value of hydrophobicity, whereas HIC contains information of possible subpopulations within the particle populations differing in surface hydrophobicity. HIC could therefore explain the observed phenomenon.

The data show the importance of having suitable characterization methods available and of characterizing particles before the performance of further experiments. The observed variations in the surface hydrophobicity are likely to be of great importance. For example when using latices for studies of phagocytosis in cell cultures, the phagocytosis will be enhanced if a batch of latex is more hydrophobic than the previous one (Müller et al., 1997). This might explain to some extent, the variations observed in cell culture studies when using a new batch of standard particles. Cell culture results are much more reproducible within one batch of particles.

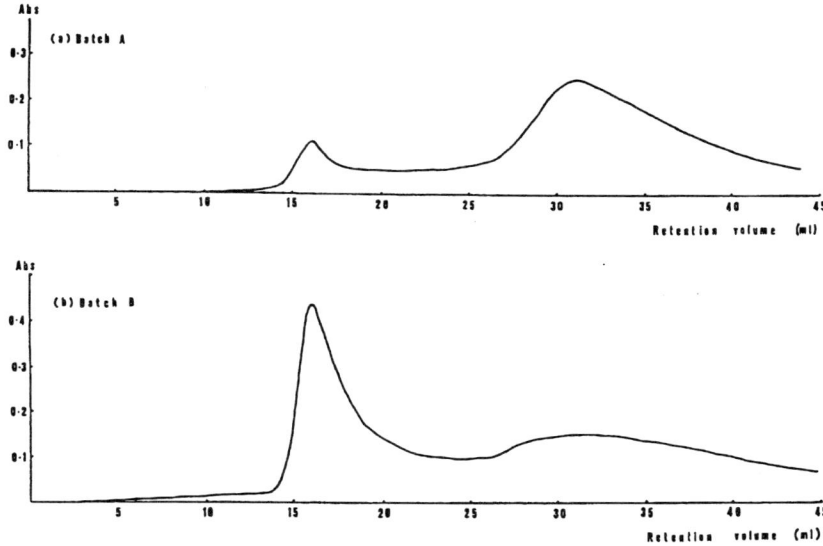

Fig. 16-4: Chromatogram of 2 batches (A and B) of fluorescently labelled carboxylated polystyrene (PSF-COOH) particles on ethyl-agarose. Elution was performed with a Triton X-100 gradient. The two batches contain a relatively hydrophilic (at 16 ml) and a more hydrophobic particle subpopulation (at 31 ml elution volume) respectively but in different proportions.

The HIC placed the particles in the same order of hydrophobicity as the Rose Bengal adsorption studies (Tab. 16-1). Binding constants and elution volumes obtained by HIC are in good agreement.

Tab. 16-1: Binding constants obtained by Rose Bengal adsorption studies and elution volumes obtained by HIC for standard latices. The sizes are photon correlation spectroscopy (PCS) data (PS - polystyrene, PSF - polystyrene fluorescently labelled; COOH, NH_2 (aliphatic), OH - functional groups introduced chemically to surface).

Particles	binding constant K (ml/µg)	elution volume (ml)	sizes
PS-0.14	0.29	32.46	140 nm
PSF-COOH batch A	0.27	31.35	213 nm
PS-AL-NH_2	0.23	30.34	212 nm
PSF	0.12	29.98	123 nm
PS-COOH	0.06	22.62	255 nm
PS-OH	0.04	20.00	220 nm

16.4 HIC of Polymeric Particles Surface-modified by Polymer Adsorption

Polystyrene latex (PS-0.14) was coated with a series of water soluble Poloxamers and with Poloxamine 908. The Rose Bengal methods are less suitable to assess the surface hydrophobicity of particles possessing adsorption layers. In general, the amount of Rose Bengal adsorbed will be extremely low if the adsorption layers are very hydrophilic (e.g. in case of Poloxamers). In addition, it cannot be excluded that a partial displacement takes place, i.e. in this case displacement of the Poloxamer / Poloxamine 908 by Rose Bengal. Therefore the hydrophobicity of the polymer coatings was investigated using HIC (Mak et al., 1988 and 1988a). All the coated particles passed down HIC columns prepared from Sepharose CL-4B without any interaction. Fig. 16-5 shows a chromatogram obtained with Poloxamine 908 coated latex whereby the elution volume of 4.5 ml is equal to the void volume of the column.

The Sepharose column was used to investigate the influence of particle size on the elution volume. Larger particles might be eluted earlier. Therefore polystyrene latex of mean size 0.9 µm (PS-0.9) was coated with Poloxamine 908 and passed down the column. No difference was found in the elution volume compared to the 0.14 µm latex (PS-0.14) and therefore, for the study a size effect could be excluded. The size difference between the investigated coated particles was relatively small (0.14 µm - 0.25 µm).

The Sepharose matrix was not suitable to resolve differences in hydrophobicity between the coating materials. Employing the more hydrophobic ethyl-agarose matrix led to a differentiation between the more hydrophobic Poloxamers (low HLB).

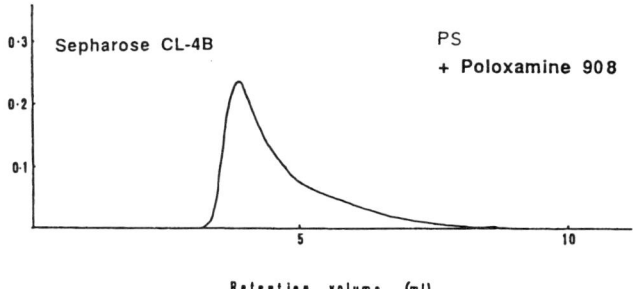

Fig. 16-5: Chromatogram of polystyrene latex particles (PS-0.14) coated with Poloxamine 908 on a Sepharose CL-4B matrix. Elution with buffered saline (0.3 M NaCl) without Triton X-100 gradient.

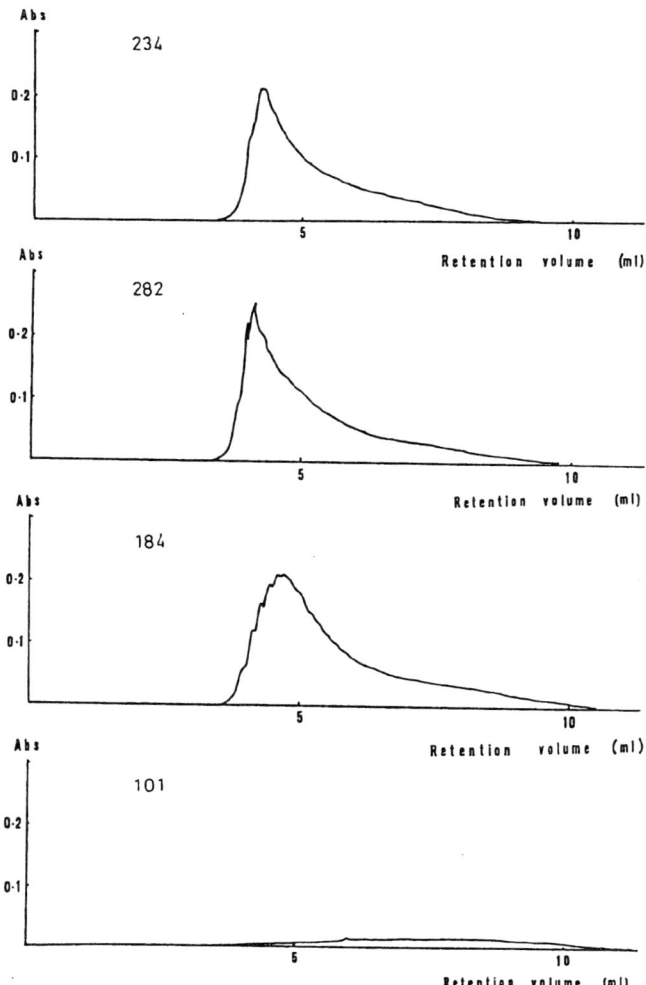

Fig. 16-6: Chromatograms of polystyrene latex particles (PS-0.14) coated with Poloxamers 234, 282, 184 and 101 on ethyl-agarose matrix. The Poloxamer coating with 101 is the most hydrophobic and the particles are mostly retained on the column.

Poloxamer 101 coated particles appeared to bind strongly to the matrix (Fig. 16-6, lowest curve), Poloxamer 184 showed a slight interaction as indicated by the fact that its elution volume was slightly higher than the void volume of the column. Poloxamer 234, 282 and the more hydrophilic polymers (Poloxamer 338 and 407, Poloxamine 908 - data not shown) passed again without interaction.

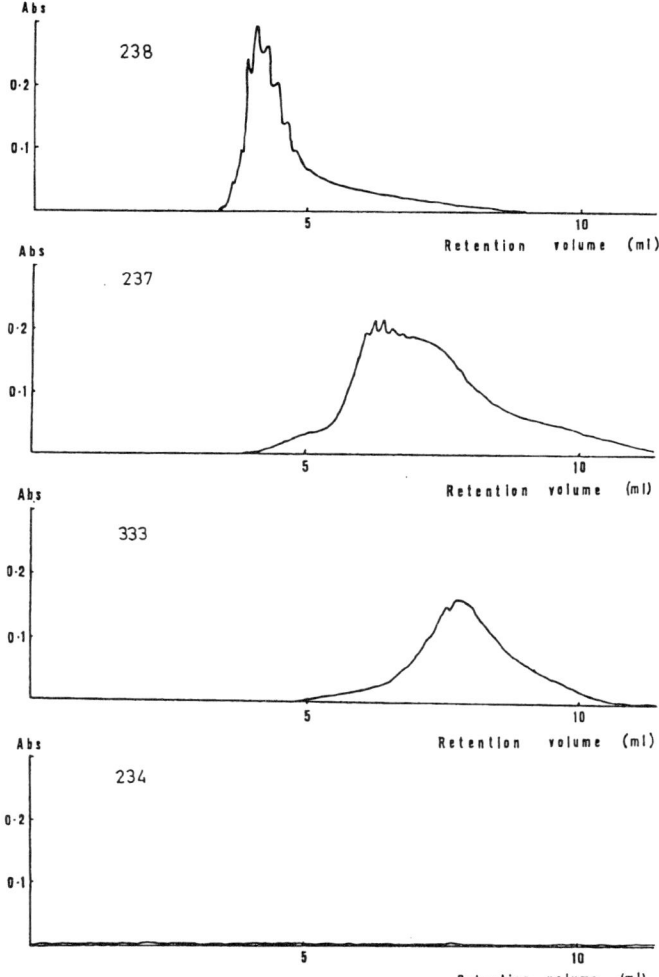

Fig. 16-7: Chromatograms of polystyrene latex particles (PS-0.14) coated with Poloxamers 238, 237, 333 and 234 on propyl-agarose matrix. The surface hydrophobicity of the coatings increases from Poloxamer 238 to 234 indicated by the increase in the elution volume. The Poloxamer 234 coated latex was too hydrophobic to be eluted and was retained on the column. Elution with buffered saline without Triton X-100 gradient.

Increasing the hydrophobicity of the matrix by using propyl-agarose led to the binding of the Poloxamer 282 and 234 coated particles (Fig. 16-7, lowest curve). The coating was too hydrophobic for the particles to elute with the void volume. A good differentiation in elution volume could be observed between the particles coated with

Poloxamers 333, 237 and 238, whereby the latter passed again without any interaction (Fig. 16-7). The propyl-agarose column proved to be suitable to place most of the Poloxamer coatings in an order of decreasing hydrophobicity (Tab. 16-2).

Tab. 16-2: Elution volumes of polystyrene latex particles (PS-0.14) coated with a series of Poloxamers and Poloxamine 908 on propyl-agarose (void volume of the column is 4.50 ml). Elution with buffered saline (0.3 M NaCl) without Triton X-100 gradient.

coating polymer	coating layer thickness (Å)	elution volume (ml)
Poloxamer 338	154	no interaction
407	119	no interaction
288	130	no interaction
238	132	no interaction
237	81	6.22
188	77	6.53
217	58	7.12
402	61	7.20
108	58	7.23
335	51	7.53
333	54	7.84
235	33	7.97
234	28	retained
184	22	retained
Poloxamine 908	133	no interaction

The latex coated with the high molecular weight (>10,000) Poloxamers 238, 288, 338 and 407 and Poloxamine 908 still passed down the column without interaction. A differentiation in the hydrophobicity of the coatings was not possible. By increasing the hydrophobicity of the matrix further using pentyl-agarose, it was hoped to detect differences in surface hydrophobicity. The latex coated with Poloxamer 338 and Poloxamine 908 again were eluted with the void volume (Fig. 16-8). However, the Poloxamer 238, 288 and 407 coated latex exhibited two peaks (Fig. 16-9). This indicates a broader distribution in surface properties whereby the first eluted peak (at around 4.5 ml) belongs to particles similar in surface hydrophobicity to those coated with Poloxamer 407 and Poloxamine 908. The second peak represents more hydrophobic particles.

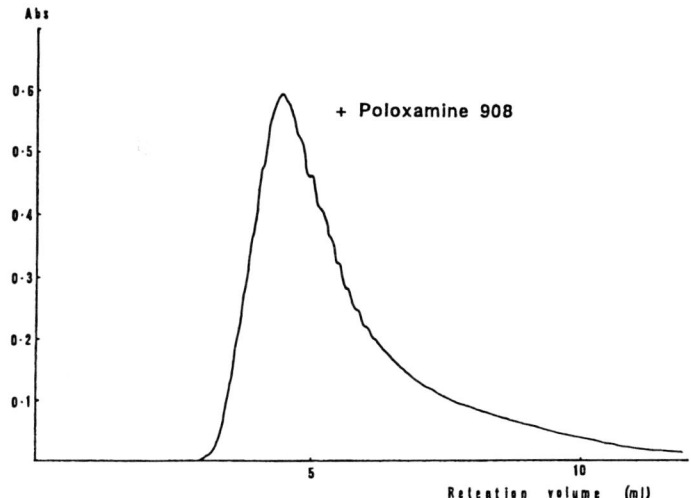

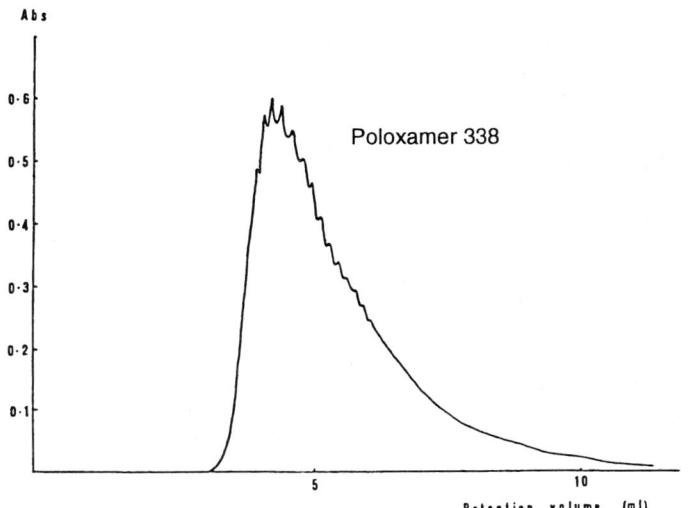

Fig. 16-8: Chromatograms of polystyrene latex particles (PS-0.14) coated with Poloxamine 908 and Poloxamer 338 on pentyl-agarose matrix. Both coated particles passed down the column without interaction.

Polystyrene particles of 0.14 μm coated with Poloxamer 407 and Poloxamine 908 passed down even a hexyl column without interaction. There were slight differences in the retention volumes for Poloxamine 908 (4.94 ml) and Poloxamer 338 (5.28 ml). In further investigations it could be shown, that a fraction of the coated particles was retained by the column. The ratio of the number of particles eluted to the number of particles bound can be used as a further characterization parameter to distinguish between such hydrophilic coatings.

HIC was also used to investigate nanoparticles surface-modified by adsorption of ethoxylated nonylphenols (e.g. Antarox™ CO and DM series). Particles modified with Antarox surfactants possessing a short ethylene oxide (EO) chain were eluted earliest, that means they appeared to be more hydrophilic than the particles modified with Antarox possessing long EO chains (Carstensen et al., 1991). This was in contradiction to the data obtained with the Poloxamer polymers. Meanwhile, it could be shown that this early elution was an artefact due to the interaction of free Antarox surfactant with the column. The free surfactant interacted with the column matrix and blocked the binding sites (Wesemeyer, 1993). To exclude this, it is necessary to remove free surfactant if the surfactant itself is rather hydrophobic (e.g. Antaraox with short EO chains), free surfactant is no problem if it is rather hydrophilic as Antarox CO 990 and DM 970 (100 and 150 EO units per chain, resp.). For Poloxamer it could be shown, that even low molecular weight poloxamers did not interact with the column (Wesemeyer, 1993). Of course, one needs to bear in mind that removal of free surfactant might destabilize a dispersion (e.g. emulsion) or change the surface properties (e.g. by partial desorption of surfactant / polymer from the particle surface). For these reasons the nanoparticles surface-modified by Poloxamers were analysed by HIC in the presence of free Poloxamer (Müller, 1991). This was possible because it could be shown that in case of Poloxamer even low molecular weight polymers do not interfere with the column.

16.5 Application of HIC to Differently Sized Nanoparticles, Emulsions and Liposomes

HIC was also successfully employed to quantify differences in surface hydrophobicity of polymeric particles coated with the same polymer but differing in particle size.

Interestingly polystyrene particles differing in size differ also in the hydrophobicity of their polymeric surface due to different polymerization conditions used for their polymerization (e.g. nature and concentration of catalysts, surfactants added as stabilizers). As a consequence, adsorption of polymers leads to adsorption layers

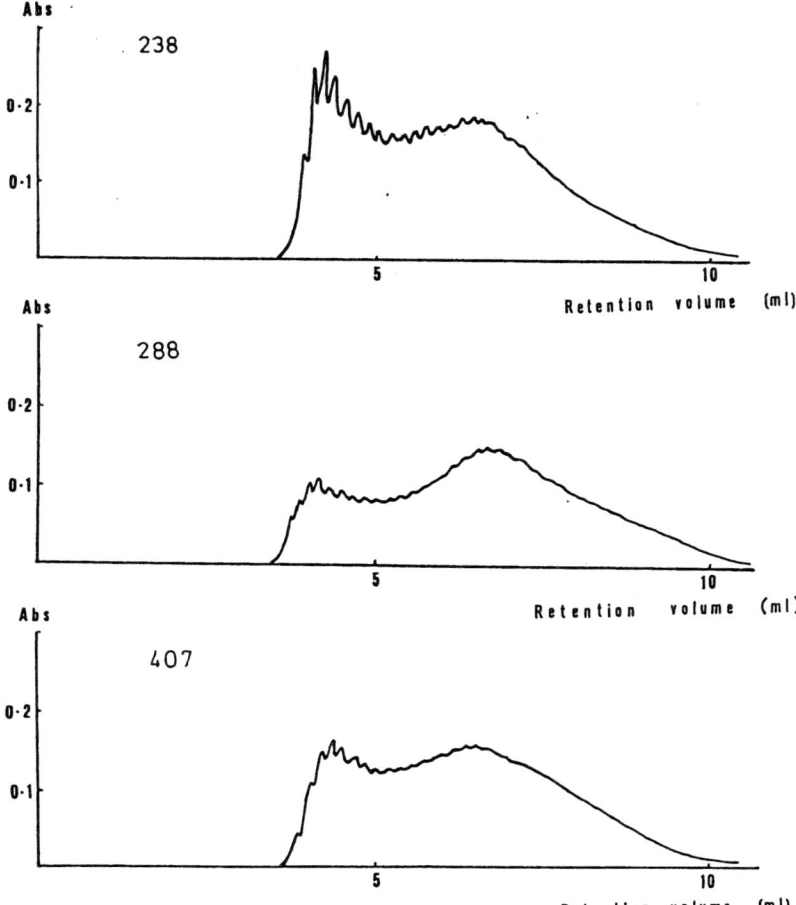

Fig. 16-9: Chromatograms of polystyrene latex particles (PS-0.14) coated with Poloxamers 238, 288 and 407 on pentyl-agarose. Two peaks were obtained. The second peak represented a particle subpopulation with a more hydrophobic surface.

differing in thickness and consequently in hydrophobicity - with important implications for their interaction with biological systems (e.g. protein adsorption after i.v. injection and resulting organ distribution). For detailed data and discussion I refer to (Müller, 1991).

HIC was also transferred to emulsions, e.g. commercial lecithin-stabilized emulsions for parenteral nutrition. Surface hydrophobicity is an important parameter determining the interaction with blood proteins and consequently uptake by the RES with possible

RES impairment (Carstensen et al., 1992). Emulsion produced with Poloxamine 908 showed similar surface hydrophobicity to Poloxamine 908-modified polystyrene particles, explaining their reduced uptake by the RES after i.v. injection (Müller, 1991).

First HIC investigations were performed with niosomes (unpublished data), depending on the composition differences in surface hydrophobicity were detectable. From this, transfer to liposomes appears to be feasable. Due to the relatively hydrophilic surface, matrix materials with a longer alkyl chain are recommended (at least propyl-agarose, i.e. similar to lecithin-stabilized emulsions).

16.6 Summary

Hydrophobic Interaction Chromatography (HIC) is a suitable method to distinguish between the surface hydrophobicity of uncoated drug carriers. This could be shown by using standard polystyrene particles with different surface properties as model carriers. A differentiation can be obtained by loading the particles on columns with alkyl-sepharose and eluting them with a Triton X-100 gradient.

The importance of such a characterization could be demonstrated by the detection of large batch to batch variations in surface properties of "standard" latex particles. Such variations can lead to reproducibility problems in investigations where surface hydrophobicity plays an important role (e.g. particle phagocytosis in cell cultures).

Coated model drug carriers were characterized by using the elution volume as a parameter, without the application of a Triton X-100 gradient. To resolve differences between very hydrophilic coatings, the hydrophobicity of the matrix was increased by using alkyl-sepharoses with increasing alkyl chain length. A series of Poloxamer coatings could be placed in order of decreasing surface hydrophobicity. Poloxamine 908 and Poloxamer 338 and 407 coatings proved to be least hydrophobic indicating their potential to reduce phagocytosis by the reticuloendothelial system.

The Rose Bengal methods give a binding constant as a measure of hydrophobicity. This is an average parameter and can only be applied on uncoated particles. HIC provides information on subpopulations with different surface hydrophobicities (resolution into different elution peaks) and can be applied to coated particles. HIC is however much more time consuming than Rose Bengal measurements (esp. the Rose Bengal partitioning method, chapter 14.3.2). A faster screening is possible by employing the so called Mini-HIC.

16.7 References

Carstensen, H., Müller, R.H. and Müller, B.W., Adsorption of ethoxylated surfactants on nanoparticles. I. Characterisation by Hydrophobic Interaction Chromatography, Int. J. Pharm. 67, 29-37 (1991)

Carstensen, H., Müller, R.H. and Müller, B.W., Particle size, surface hydrophobicity and interaction with serum of parenteral fat emulsions as parameters related to RES uptake, Clinical Nutrition 11, 289-297 (1992)

Mak, E., Davis, S.S., Illum, L. and Müller, R.H., Determination of surface properties of 'standard' latex particles, Abstr. British Pharmaceutical Conference, 22-24 September Jersey, 100P (1986)

Mak, E., Müller, R.H., Davis, S.S. and Illum, L., Hydrophobic Interaction Chromatography and Laser Doppler Anemometry for the characterisation of coated drug carriers, Acta Pharm. Technol. 34, 23S (1988)

Mak, E., Müller, R.H., Davis, S.S. and Illum, L., Characterisation of colloidal carriers for drug targeting, Acta Pharm. Technol. 34, 20S (1988a)

Müller, R.H., Davis, S.S., Illum, L. and Mak, E., Surface characterisation of colloidal drug carriers coated with polymers, Macromolecular Preprints, 161 (1986)

Müller, R.H., Rühl, D., Lück, M. and Paulke, B.-R., Influence of fluorescent labelling of polystyrene particles on phagocytic uptake, surface hydrophobicity and plasma protein adsorption, Pharm Res., in press

Müller, R.H., Colloidal Carriers for Controlled Drug Delivery and Targeting - Modification, Characterisation and In Vivo Distribution, Wissenschaftliche Verlagsgesellschaft Stuttgart and CRC Press Boca Raton, 1991

Pahlmann, S., Rosengren, J. and Hjerten, S., Hydrophobic Interaction Chromatography on uncharged Sepharose derivatives. Effects of neutral salts on the adsorption of proteins, J. Chromatogr. 131, 99-108 (1977)

von Hippel, P.H. and Schleich, T., The effects of neutral salts on the structure and conformational stability of macromolecules in solution. In Structure and Stability of Biological Macromolecules (Timasheff, S.N., Fasman, G.D., eds.) Marcel Dekker, New York, 417-574 (1969)

Wesemeyer, H., Mikrokalorimetrische und chromatographische Untersuchungen zur Oberflächenmodifikation von partikulären Arzneistoffträgern durch Tenside, Dissertation, Christian-Albrechts-Universität zu Kiel (1993)

Address of the author:
Prof. Dr. R.H. Müller
Department of Pharmaceutics, Biopharmaceutics & Biotechnology
Free University of Berlin
Kelchstr. 31
D-12169 Berlin

17 HIC of Iron Oxide Dispersions Stabilized by Different Macromolecules

Dr. K. Thode, Dr. M. Kresse, Prof. Dr. R.H. Müller, Berlin

17.1 Introduction

Iron oxide particles are used as contrast agents in magnetic resonance imaging (MRI). In aqueous dispersions these iron oxides are stabilized by macromolecules, the diameter of the iron oxide core is appr. 4 - 8 nm, the particle diameter including surrounding macromolecules depends on the synthesis parameters.

The compositions of iron oxide dispersions which were analyzed by HIC is shown in Tab. 17-1. Looking at the data, it is not difficult to decide which type of alkyl-agarose would be potentially most suitable for interaction with the iron oxide particles. The batches are rather hydrophilic because of their hydrophilic stabilizing macromolecules. So the column material ought to be highly hydrophobic to bind these particles. Furthermore, the samples can be adjusted to possess same contents of Fe, so that it is a reasonable possibility for comparison of the surface hydrophobicity.

Tab. 17-1: Composition, PCS size and zeta potential of iron oxide dispersions (CDx = carboxydextran, appr. 2.3 kDa; Dx = dextran, appr. 1 kDa; CSA = chondroitin-4-sulfate, appr. 40 kDa)

sample	A	B	C	D	E	F
Fe-content in % (w/v)	5.8	5.4	5.7	3.3	1.7	1.7
coat in % (w/v)	6.7% CDx	5.6% CDx	5.2% CDx	1.6% CDx	0.8% CDx 19.3% Dx	3.3 % CSA
core	Fe_2O_3	Fe_2O_3	Fe_3O_4	Fe_2O_3 / Fe_3O_4	Fe_2O_3 / Fe_3O_4	Fe_2O_3 / Fe_3O_4
∅ [nm]	65.4	60.1	31.2	38.4	43.5	105.3
zetapotential [mV] in distilled water	-45.8	-45.4	-32.2	-32.3	-32.7	-54.6

Compared to HIC, there is no on-line detection possible in Mini-HIC; to quantify the portion of bound and unbound particles the eluate is collected in fractions of appr. 333 mg (by placing collecting cuvettes on balance). The application of UV-VIS spectroscopy for iron oxide particles and comparable samples is shown step by step.

17.2 Choice of the Appropriate Wavelength

Choosing a suitable wavelength is important because the absorption has to be due mainly to the iron oxide particles. Disturbances such as the absorption of excess macromolecules or other substances should be avoided, minimized or removed by mathematical calculation (see below). Further, the concentration of the iron oxide dispersion has to be chosen that an absorption value of about 1.0 is not exceeded - a total elution in one fraction assumed. Of course, this concentration must not lead to an overload of the column material.

First, spectra of the dispersions and the associated macromolecule solutions were scanned (Uvikon 940, Kontron Instruments). This should detect the degree of overlay of both absorption curves and allow to choose a wavelength with a low macromolecule absorption. Fig. 17-1 shows the spectrum of carboxydextran (left) and the iron oxide dispersion C (right) in the range of 200 - 450 nm. The presentation of the absorption at higher wavelengths is not necessary because of very low absorption.

The macromolecule solution showed an absorption peak at 274 nm whereas sample C showed a decreasing absorption with increasing wavelength without having any maximum. Regarding the left figure, solutions containing other macromolecules than

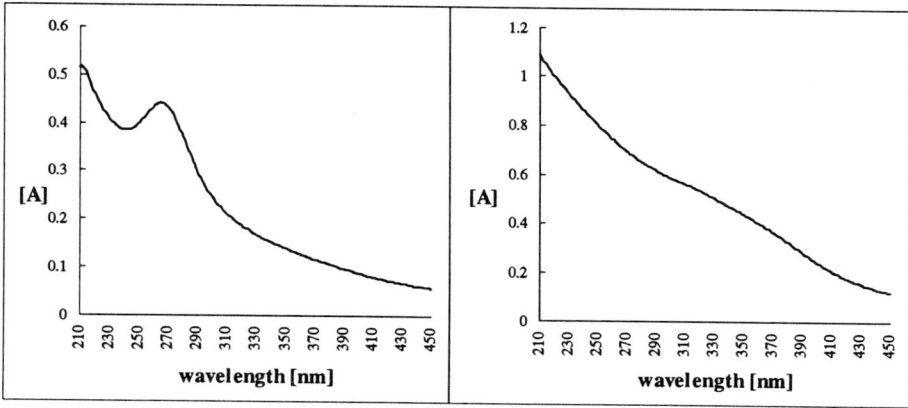

Fig. 17-1: Wavelength scans of a carboxydextran solution (0.1% [w/v], left) and of sample C (diluted by factor 5,000, right) in the range of 200 - 450 nm

CDx (mixture of CDx and Dx; CSA) gave comparable spectra. All the iron oxide dispersion samples provided similar spectra, as shown on the right. Taking the lower concentration of CDx in sample C into account, it is clear, that the absorption is mainly based on the iron oxide particles (factor of 5).

When dividing the absorption values of the iron oxide dispersion by the absorption values of the macromolecule solution in the range of 190 to 700 nm, the calculated quotient is a measure to which extent the absorption is dominated by the particles. A high absorption of iron oxide particles at a specific wavelength leads to a high quotient.

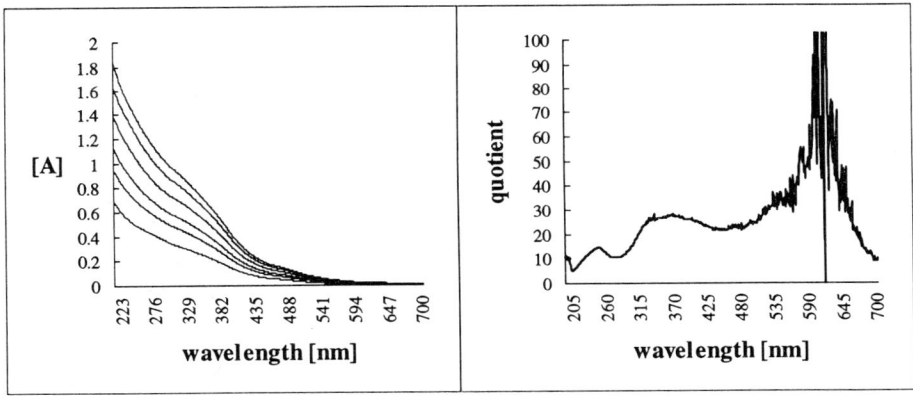

Fig. 17-2: Left: wavelength scan of sample E in concentrations ranging from 0.04% [w/v] to 0.14% [w/v] of the original sample in steps of 0.02%; right: ratio of absorption for sample E. The absorption values of the diluted dispersion E (0.1% [w/v]) are divided by the absorption values of the macromolecule solution (0.2% [w/v]) in the range of 190 - 700 nm

In order to receive significant results the concentration of the macromolecule solution was ten times higher than its original value in sample E (0.2% instead of 0.02%). Scanning the genuine concentration, as it was in sample E, would have led to low absorption values. From Fig. 17-2 it was not difficult to choose an appropriate wavelength. The highest quotients were in the range of 550 - 650 nm. It was not possible to measure in this region because the concentration of the dispersion would not have been sufficient to receive a satisfactory absorption value. The application of a larger volume would surely overload the agarose matrix. Therefore the iron oxide dispersions were determined quantitatively at 350 nm implying a sufficiently high absorption of particles which was only little overlaid by macromolecule absorption.

The very low absorption of the macromolecules can additionally be calculated as follows:

The total absorption of an iron oxide dispersion consists of the particle absorption and the macromolecule absorption according to:

$$A_{350} = A_{350}(\text{iron oxide}) + A_{350}(\text{macromolecule}) \qquad \text{Eq. 17-1}$$

Each contribution can be calculated via concentration and specific absorption:

$$A_{350}(\text{iron oxide}) = [\text{iron oxide}] \cdot \text{spec. } AC_{350}(\text{iron oxide}) \qquad \text{Eq. 17-2}$$

$$A_{350}(\text{macromolecule}) = [\text{macromolecule}] \cdot \text{spec. } AC_{350}(\text{macromolecule}) \qquad \text{Eq. 17-3}$$

A_{350} = absorption at 350 nm
[ ] = concentration of iron oxide or macromolecule, respectively
spec. AC_{350} = specific absorption coefficient at 350 nm

It is possible to determine A_{350}(macromolecule) directly and to obtain A_{350}(iron oxide). Measuring the absorption of macromolecule solutions with increasing concentrations at 350 nm enables to fit a calibration straight line (Fig. 17-3). The specific absorption coefficient of the macromolecule solution, calculated as slope of the straight line, in multiplication with the macromolecule concentration results in A_{350}(macromolecule) (cf. Eq. 17-3). Eq. 17-1 can then be solved. An example makes this clear:
An absorption of 0.430 was measured for a diluted sample of dispersion B (0.2% [w/v], i.e. original dispersion diluted by factor 500) at 350 nm. The calibration straight line of the macromolecule carboxydextran is shown in Fig. 17-3.

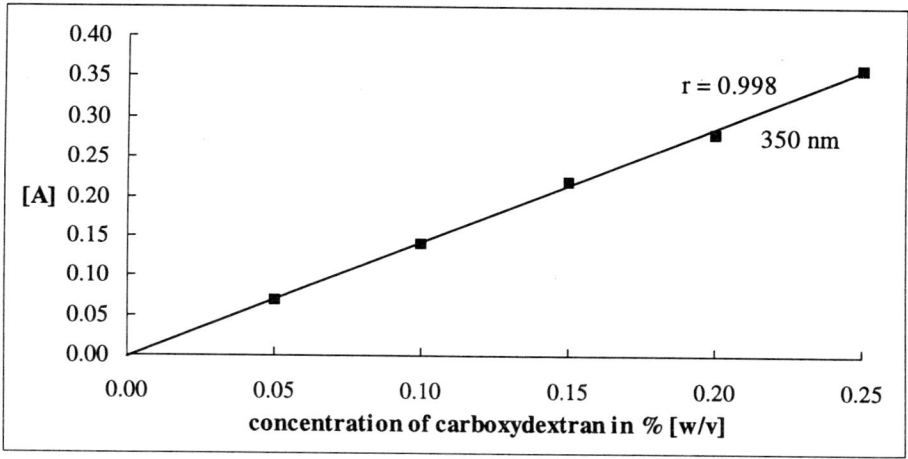

Fig. 17-3: Calibration line of carboxydextran at 350 nm in distilled water.

From this, spec. AC_{350}(macromolecule) could be determined as 1.44. The concentration of carboxydextran was 0.0012% [w/v] in the diluted sample of dispersion B (0.2%). Using Eq. 17-1 it can be calculated:

$$0.430 = A_{350}(\text{iron oxide}) + 0.0012 \cdot 1.44 \qquad \text{Eq. 17-4}$$

This means: at 350 nm 99.6% of the total absorption of sample B is caused by the iron oxide particles. The macromolecule contribution is within the error of the method and can be neglected in this example.

17.3 Running Mini-HIC of Iron Oxide Dispersions

Mini-HIC was performed as described above. The investigations consisted of:
1. Screening of hydrophobic interaction using ethyl-, butyl- and hexyl-agarose; elution with phosphate buffer pH 6.8, followed by elution with a buffered solution (1.0%) of Triton X-100 (1.0 ml for each fraction, respectively). The content of particles in the fractions was quantified by UV-VIS spectroscopy.
2. Repetition of Mini-HIC using the alkyl-agarose that has shown most interaction with iron oxide particles; elution with phosphate buffer in smaller fractions (appr. 0.333 ml (!) instead of 1.0 ml for each fraction, determination by weighing) to yield a more detailed picture. The content of particles in the fractions was quantified by UV-VIS spectroscopy.
3. Optionally, the particle size of each fraction may be determined. Thus, further information concerning the relationship between particle size and surface hydrophobicity can be obtained. In addition, artefacts due to aggregation of particles can be detected (cave! Aggregates yield higher absorption values!).

17.3.1 Screening

It was observed that ethyl- and butyl-agarose were not hydrophobic enough to interact with iron oxide particles. No particles were bound on the ethyl-agarose column, little of the samples A, B and F on the butyl-agarose column (Fig. 17-4, left and middle). Considering the precision of this screening method (1.0 ml was collected by a visual mark in the cuvette), the small differences on the butyl-agarose were not significant. In contrast to this, differences in the surface hydrophobicity were detected using hexyl-agarose as column material (Fig. 17-4, right).

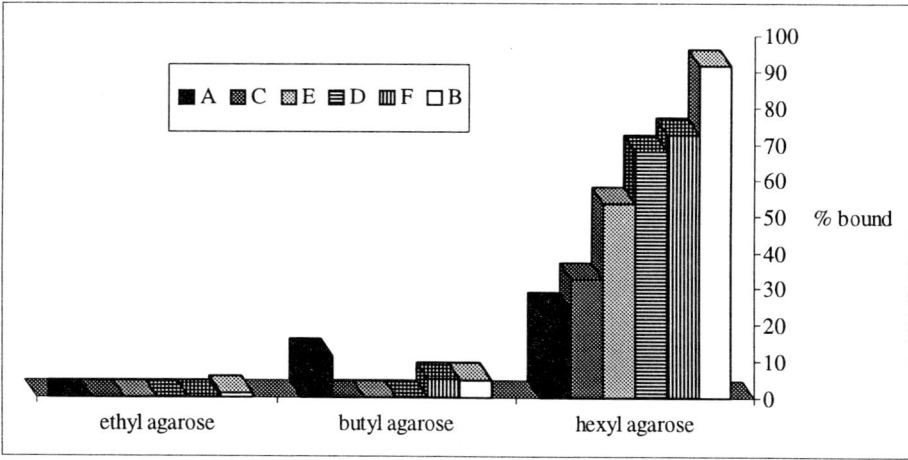

Fig. 17-4: Percentage of the iron oxide dispersions A - F bound on different agaroses varying in the length of alkyl ligands (from screening procedure, 1.0 ml (!) fractions)

Triton X-100 solution (1.0%) was not able to wash off any iron oxide particles. These were bound too strongly to the hexyl-agarose matrix, i.e. they were irreversibly bound. Carrying out further investigations with the samples A - F allowed determination of the reproducibility at about ± 5%.

17.3.2 Repetition of Mini-HIC with Reduced Volumes per Fraction

In order to obtain more detailed information concerning the elution behaviour Mini-HIC was repeated using appr. 333 mg per fraction (cf. Fig. 17-5).
The high resolution Mini-HIC revealed that the dispersions did not only differ in the amount eluted and irreversibly bound to the hydrophobic matrix but also in their elution profile. For most samples, the first fraction eluted comprised more of the iron oxide particles than the others. However, sample E, and especially sample C, showed the highest amount of particles being eluted in the second fraction. This is caused by a stronger retardation for these obviously more hydrophobic particles. Sample E contained a certain percentage of particles being strongly retarded by hexyl-agarose, therefore it was more volume needed to elute unbound particles and they appeared enriched. This behaviour was not shown by other dispersions and was attributed to the complex composition of this sample (CDx and Dx).

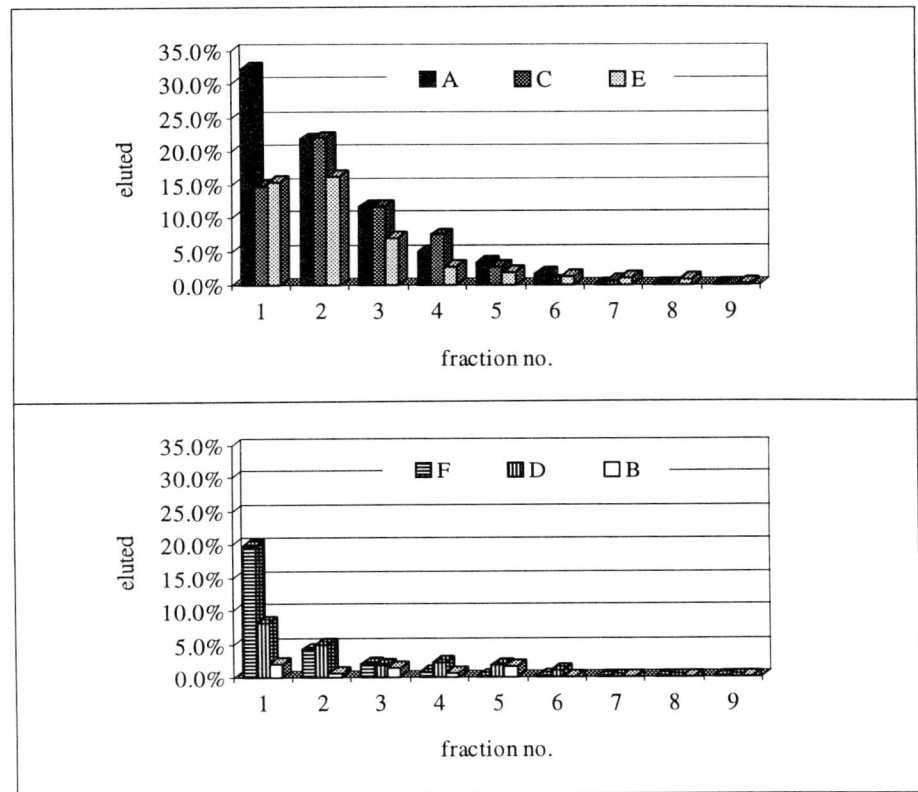

Fig. 17-5: Percentage of the iron oxide dispersions A, C, E (upper) and F, D, B (lower) eluted from hexyl-agarose. Elution was performed using phosphate buffer pH 6.8 in 0.333 ml fractions (fraction eluted is expressed in % of 100% sample loaded on the column; the difference between the total percentage eluted in the 9 fractions and 100% is the percentage of the dispersion irreversibly bound).

17.3.3 Particle Size in Eluted Fractions from Mini-HIC

As a matter of fact an iron oxide dispersion does not exist of particles having exactly the same properties, e.g. surface hydrophobicity. That means they show a certain degree of polydispersity in their surface properties. By using Mini-HIC it is possible to separate fractions of iron oxide dispersions differing in surface hydrophobicity. Changes in the surface hydrophobicity may be correlated to the particle diameter. In

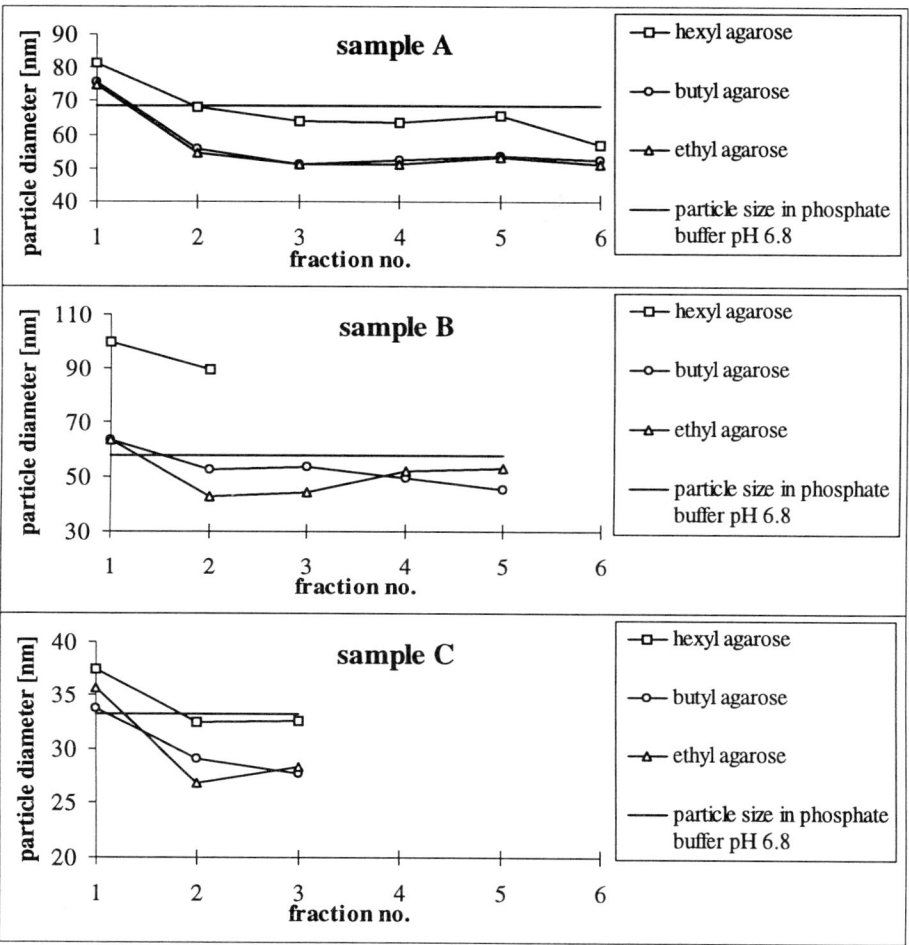

Fig. 17-6: Particle diameter of different fractions of dispersions A (top), B (middle) and C (bottom) eluted over ethyl-, butyl- and hexyl-agarose (straight line parallel to x-axis: mean diameter of each dispersion).

order to show a possible correlation between surface hydrophobicity and particle size, the diameter of the particles in each fraction was determined using a Zetasizer 4 (Malvern Instruments, UK). Due to the small volume of a Mini-HIC fraction being eluted (about 333 µl) as well as the very low particle concentration in later fractions, particle sizing was not feasible for all fractions, as can be seen in Fig. 17-6.

For all investigated samples, the first elution fraction over hexyl-agarose clearly contained the largest particles, partly 35 nm larger than the average diameter (sample B). In general, elution of particles in later fractions revealed a decrease in particle size. Due to the small bed volume of the Mini-HIC and previous size effect studies on elution (cf. chapter 16) it could be excluded that the elution of the larger particles in the first fraction is a size effect. Therefore, the smaller particles must be more hydrophobic leading to an increased interaction with the hydrophobic matrix. Regarding the larger particles, it is assumed that the macromolecule carboxydextran is fixed on the iron oxide in multilayers leading to an increased surface hydrophilicity. Thus, it was possible to separate a subpopulation of the dispersion having more hydrophilic properties and larger diameters than the average dispersion.

17.4 Conclusions

It could be demonstrated that Mini-HIC is a suitable method to separate fractions of dispersions differing in surface hydrophobicity. This separation can be used to characterize particle populations. HIC on large columns can be employed for preparative separation. This opens e.g. the perspective to investigate the in vivo fate of particle subpopulations differing in their properties.

Address of the corresponding authors:
Dr. K. Thode, Prof. Dr. Rainer H. Müller
Department of Pharmaceutics, Biopharmaceutics & Biotechnology
Free University of Berlin
Kelchstr. 31
D-12169 Berlin

18 Two-Dimensional Polyacrylamide Gel Electrophoresis (2-D PAGE) for the Analysis of Protein Adsorption onto Surfaces

Dr. T. Blunk, Massachusetts Institute of Technology, Cambridge, USA

18.1 Why 2-D PAGE for the Analysis of Protein Adsorption?

18.1.1 Protein Adsorption Influences the Success of Biomaterials

In many biomedical applications, where new materials are under investigation, the adsorption of plasma proteins is crucial for the success of the materials involved. The biocompatibility of clinical implants, artificial organs or tissues, and intravenously injectable colloidal drug carriers can be influenced by preferentially adsorbed plasma proteins [1-3]. On the one hand, for example, the success of artificial organs or tissues can be limited by the initiation of blood coagulation and complement activation. On the other hand, it may mediate favourably the adhesion of desirable mammalian cells. For colloidal drug carriers, adsorbed proteins may either allow the recognition by the mononuclear phagocytic system or shield the carriers from MPS recognition and mediate the uptake by specific target cells.

18.1.2 Towards the Investigation into Biocompatibility

If the biocompatibility of materials is to be investigated, it must be decided which type of study makes the most sense for a particular application.

Studies of protein adsorption from buffers containing single proteins or mixtures of a few proteins are useful for the determination of general adsorption principles. Protein adsorption to surfaces from complex protein mixtures like blood or plasma, however, cannot be satisfactorily modelled by those studies. Instead blood-material interactions have to be investigated in plasma or blood itself [2].

There are two major approaches for the analysis of protein adsorption from plasma. In the first one, purified radiolabeled proteins are added in small amounts to the plasma as tracers. This approach gives quantitative data on the adsorption of individual proteins. Thus it is appropriate, if the researcher is interested very specifically in a particular protein, but it is inefficient, if one is interested in a more comprehensive qualitative answer. Certainly, by just examining individual proteins, others will be missed, which may be important for the desired application.

The second approach is more qualitative in nature and targets the identification of the multitude of proteins adsorbed on a surface. After plasma contact, the proteins are eluted from the surface and subsequently separated by electrophoresis or other techniques. Commonly used so far is the SDS-polyacrylamide gel electrophoresis (SDS-PAGE), which separates the proteins according to their molecular weight. This analysis, however, results in bands of proteins from which it still cannot be determined which specific proteins were adsorbed on the surface. This requires again the testing for specific proteins, for example using immunoassays. Again it is very likely that many of the proteins remain unidentified [2].

The *two-dimensional* polyacrylamide gel electrophoresis (2-D PAGE), however, does not suffer from these limitations. 2-D PAGE is a combination of two different electrophoretic techniques, that is isoelectric focusing (IEF) and SDS-PAGE. Therefore the proteins are resolved according to two independent parameters, namely charge and size. This allows 2-D PAGE to separate a mixture of proteins into *single components* with the highest resolution available. It is therefore the best technique for the simultaneous separation and identification of complex protein mixtures. In addition, in combination with the silver-staining technique, it is capable of providing semi-quantitative measurements of amounts of protein [4, 5].

18.1.3 2-D PAGE in Drug Targeting with Colloidal Carriers

After intravenous injection colloidal drug carriers immediately interact with plasma proteins. The adsorbed protein pattern depends on the surface properties of the drug carriers. In turn the adsorbed plasma proteins determine the *in vivo* behavior of the drug carriers ("differential adsorption") [3, 6]. Currently there is still little knowledge about the importance of *particular* proteins for an effective drug targeting. But it is evident that 2-D PAGE can play a vital role in the identification of key proteins and in the establishment of correlations of the surface characteristics with the adsorbed protein pattern and, in turn, the correlations of the adsorbed proteins with the *in vivo* distribution of the drug carriers [7].

18.2 How to Do the Analysis?

18.2.1 Overview

This section will demonstrate how to perform the analysis of adsorbed proteins using 2-D PAGE, with potential colloidal drug carriers serving as an example. The electrophoresis technique itself, which was first successfully applied to protein analysis over 20 years ago [8], will be briefly reviewed. The main focus, however, will be on the sample preparation for the electrophoresis, since this is of key importance for the analysis of *adsorbed* proteins.

The sample preparation involves the incubation of the drug carriers in plasma, the separation of the carriers from the plasma, and the desorption of the adsorbed proteins from the drug carriers. The 2-D PAGE itself consists of an isoelectric focusing in the first dimension, followed by a SDS-PAGE in the second dimension. After that the gels are silver-stained, scanned with a laser-densitometer and the data can be processed with software especially designed for the analysis of 2-D PAGE gels [9, 10].

18.2.2 2-D PAGE and Silver Staining

The isoelectric focusing basically is an electrophoresis within a pH gradient [4]. Amphoteric molecules, in this case the proteins, migrate to that position in the gel at which the pH of the gel equals the isoelectric point of the protein. Since gels with large pores are employed, size is no restriction to migration here. Thus the proteins are separated solely based on their charge. Most of the work so far has been done with carrier ampholytes generating the pH gradient. A more elegant method employs immobilized pH gradients [11]. With the carrier ampholytes, the separation in the first dimension takes place in tube IEF gels (Tab. 18-1) [12].

After the IEF, the proteins, which are still within the tube gel, are coated with sodium dodecyl sulfate, which results in a constant charge to mass ratio of the coated proteins. Then the tube gels are laid on top of one edge of the slab gels of the second dimension, into which the proteins migrate during the subsequent electrophoresis, the SDS-PAGE [4]. Since the original charge differences are masked in the SDS-PAGE and a linear acrylamide gradient creates a restrictive gel, the proteins are resolved only according to their molecular weight (Tab. 18-1) [12]. The 2-D PAGE results in a polyacrylamide gel in which each particular protein or polypeptide spot has its own characteristic coordinates [13, 14].

After the electrophoresis the slab gels are stained with silver. Silver ions are added to the gel. The proteins complex with the silver ions and the stain can be developed by reduction to metallic silver [5]. Silver staining is very sensitive with a detection limit as low as 0.05 - 0.1 ng protein / mm^2 [15] which makes it a powerful detection method suiting the high resolution of the 2-D PAGE. A linear relationship between the silver staining density and the protein concentration was found over a 30-fold [16] to 40-fold [15] range, but one has to be aware of that this linear range is not infinite and especially that the slope is characteristic for each protein. Therefore in quantitative inter-gel comparative studies the comparison has to be limited to homologous protein spots on each gel. In other words, the data from silver staining, when different proteins are compared, can only be semi-quantitative [17].

Tab. 18-1: Features of 2-D PAGE Protocol [after 12]

IEF
* Tube gels: Carrier ampholytes, gradient pH 4-8.
*   4 % T, 2.6 % C (PDA as crosslinker)
size: 160 x 1.5 mm.
* Focusing: 200 V for 2 h, 500 V for 5 h, 1000 V for 10.5 h (13400 Vh).
SDS-PAGE
* Slab gels: linear acrylamide gradient
9-16 % T, 2.6 % C (PDA as crosslinker)
size: 160 x 160 x 1.5 mm
*   Running conditions: 40 mA /gel (~ 4.5 h)

18.2.3 Sample Preparation for the Analysis of *Adsorbed* Proteins

The standard protocol of the sample preparation is shown in Tab. 18-2a [7, 18]. As the first step the particles are incubated with human plasma in Eppendorf tubes at 37 °C. The incubation time routinely chosen is 5 min, but this can be varied according to the type of study, e.g. analysis of kinetics of adsorption. One important variable in the incubation process is the concentration of the plasma in which the particles are incubated, an aspect which is sometimes overlooked in the literature in adsorption studies employing SDS-PAGE. In Fig. 18-1 the detected protein amounts on polystyrene model particles are compared after adsorption fom 66 % and 98 % plasma. There were considerable differences noticeable. While the amount of fibrinogen decreased with increasing plasma concentration, the amount of PLS:6 increased drastically. Also a considerable difference was noticable for $\alpha 2$-macroglobulin. This illustrates that the plasma concentration should be carefully considered before designing an experiment.

The next step in the sample preparation is the separation of the colloidal particles from the plasma, which is routinely done using centrifugation. The centrifugation conditions may be altered according to the size and density of the particles, but 15000 g for 60 min is often used. The supernatant is discarded and the pellet redispersed in distilled water. Centrifugation and redispersion are repeated four times.

The last part in the sample preparation is the desorption of the proteins from the carriers. After the supernatant from the centrifugation is discarded for the last time, the particles are redispersed in a solution containing SDS and dithioerythritol and incubated at 95°C for 5 min. Then the suspension is cooled to room temperature and a solution is added which contains CHAPS, DTE, urea and ampholytes. CHAPS is a

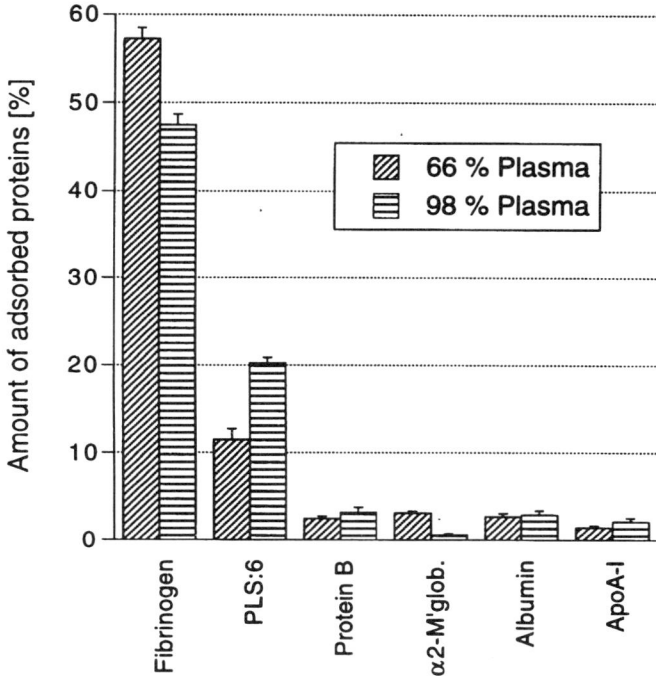

Fig. 18-1: Influence of plasma concentration on the protein adsorption pattern. Protein adsorption from plasma with different concentrations onto polystyrene model particles (diameter 200 nm). Amounts of single proteins are expressed as percentages of the overall adsorbed protein amounts (Percentages represent the mean of two experiments, error bars = standard deviation) - PLS:6 (13) and Protein B (7) are unidentified proteins.

solution of the detergents applied in the desorption process must desorb proteins from the surface *and* must be compatible with the subsequent isoelectric focusing. Both requirements are fulfilled using the compounds in the presented protocol. In general, the protein amount desorbed from the particles is greater than 90 %, as determined by the BCA protein assay [7, 19].

An alternative protocol uses filtration instead of centrifugation as the separation method (Tab. 18-2b) [7, 20]. After incubtion of the particles, the suspension is immediately filtered through a polycarbonate filter, for example with a pore size of

Tab. 18-2: Sample Preparation for the Analysis of Adsorbed Proteins

---

a) Standard protocol (Centrifugation)
1. Incubate particles with human plasma in Eppendorf tube at 37°C for 5 min.
2. Centrifuge, 15000 g, 60 min.
3. Discard supernatant, redisperse particle pellet with distilled water.
4. Repeat steps 2. and 3. four times.
5. After discarding of supernatant for the last time, redisperse particles in solution containing SDS and DTE. Incubate at 95°C for 5 min.
6. Cool to room temperature. Add solution containing CHAPS, DTE, urea and ampholytes.
7. Apply to IEF, process like normal 2-D PAGE plasma samples.

b) Alternative protocol (Filtration)
1. see a) step1.
2. Filter suspension through polycarbonate filter, e.g. 0.1 µm.
3. Wash particles on filter with distilled water.
4. Remove particles from filter using a small spatula.
5. Redisperse particles in solution containing SDS and DTE. Incubate at 95°C for 5 min.
6. see a) step 6.
7. see a) step 7.

---

100 nm. The particles on the filter are thoroughly washed with distilled water and the particles are removed from the filter using a small spatula. After that the particles are processed as in the standard protocol. In Fig. 18-2 the amounts of adsorbed proteins on polystyrene model particles are shown. The example demonstrates that no considerable differences were detected after the two different separation methods had been employed.

Both methods give reproducible and comparable results. No appreciable protein losses occur during centrifugation steps or due to adsorption on the filter. But there are several criterions which are important for the decision to be made (Tab. 18-3). One has to be aware of the fact that only with the centrifugation method a quantitative recovery of the particles can be achieved. In the process of removing particles from the filter there is always a small fraction retained on the filter. Also the analysis of very small particles, with a diameter smaller than 100 nm, is only possible using the centrifugation. The filtration is made impossible by small pore sizes because of blocking when filtering highly concentrated plasma and because of the retention of larger proteins. Inspite of these disadvantages, the filtration is beneficial in several

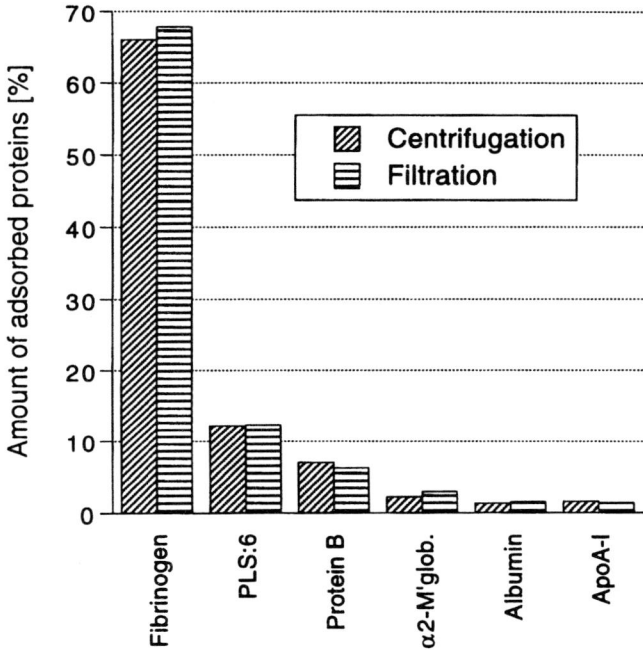

Fig. 18-2: Comparison of detected proteins after two different separation methods of particles from plasma during the sample preparation: Amounts of single proteins are expressed as percentages of the overall adsorbed protein amounts - PLS:6 (13) and Protein B (7) are unidentified proteins.

Tab. 18-3: Separation of the Drug Carriers from Plasma - Centrifugation *vs* Filtration Both methods give reproducible and comparable results. No protein losses occur during centrifugation steps or due to adsorption on the filter.

	Centrifugation	Filtration
Quantitative recovery of particles	yes	no
Analysis of small particles (<100 nm)	possible	impossible
Kinetics of early stages of adsorption	not measurable	measurable
Speed	slow	fast

circumstances. The major advantage is the possibility to measure the early stages of the kinetics of protein adsorption. The incubation can be terminated within seconds so that the adsorbed proteins after only seconds of incubation become detectable. In contrast, the centrifugation method permits additional contact time during centrifugation. Another advantage of the filtration is the decreased time required for separation of most particles. In conclusion, it is possible to choose the appropriate separation method according to the needs of a particular experiment.

18.3 What to Expect from 2-D PAGE? - A Study of Kinetics of Protein Adsorption as an Example

18.3.1 Introduction

The results of a study dealing with the kinetics of protein adsorption on colloidal drug carriers will serve as an example in order to demonstrate the capability of 2-D PAGE.

The *in vivo* behavior of i.v. injected drug carriers is usually studied over a period of minutes and sometimes as long as days. Protein adsorption onto solid surfaces has been reported to be time-dependent, including the displacement of initially adsorbed proteins by others [2, 21-24]. Consequently, while circulating in the blood, a change in adsorption patterns on drug carriers is likely to occur.

The investigation of protein adsorption kinetics is closely related with the name of Leo Vroman. From his experiments, Vroman has postulated a rapid sequence of protein adsorption and displacement on solid surfaces, where more abundant proteins are displaced by less abundant [21, 22]. This phenomenon, especially for transiently adsorbed fibrinogen, is referred to as the "Vroman-Effect". Its extent depends on the surface to which the proteins are adsorbed [2].

The displacement often occurs within seconds or even within fractions of a second. Therefore, the initially adsorbed proteins are sometimes not detected at all, due to limitations of the experimental technique. This especially applies to albumin, which is by far the most abundant protein in plasma.

However, if certain proteins are adsorbed only transiently due to displacement by others, it seems likely that by diluting the plasma sufficiently, the concentration of the displacing proteins would be decreased to such an extent, that the residence times of the proteins which were adsorbed first would be prolonged. Thus these proteins would become detectable [2]. In other words, the amount of a single plasma protein adsorbed onto a surface detected after a specific incubation time is a function of the plasma dilution. Therefore, to study the very early stages of adsorption onto model particles, the adsorption from diluted plasma was analyzed.

18.3.2 Early Stages of Protein Adsorption

Polystyrene particles with a diameter of 1 µm which were surface-modified by adsorption of poloxamer 407 were used as model particles. Plasma was diluted with distilled water, adsorption was performed from plasma with a concentration of 0.08 %, 0.8 % and 80 % plasma for 5 min each. Fig. 18-3 shows close-ups of the 2-D PAGE gels.

The three gels show large qualitative differences, which clearly illustrate the principle of early adsorption and displacement. On the adsorption pattern from 0.08 % plasma albumin was present in the highest amount. In addition, there were a considerable amount of transferrin and smaller fractions of fibrinogen and IgG detected. On the gel which resulted from adsorption from 0.8 % plasma, fibrinogen was the dominant protein. The fraction of albumin was strongly reduced as compared to the adsorption pattern from 0.08 % plasma. Instead, plasminogen and apolipoprotein C-III (not shown) were detected in small amounts. On the adsorption pattern from 80 % plasma ApoC-III became one of the major proteins together with PLS:6 (unidentified protein, named after [13]) and apolipoprotein J, fibrinogen was decreased to small amounts.

What do these results mean for the adsorption kinetics in highly concentrated plasma? The three adsorption patterns resulted from an incubation time of five minutes each. Fibrinogen seemed indeed to have been adsorbed transiently on the particles. Under the conditions of adsorption from 80 % plasma fibrinogen had already been displaced by others (PLS:6, apoJ, apoC-III) within the time frame of incubation. However, by sufficiently diluting the plasma (0.8 %), the concentration of displacing species was decreased below effective levels, so that after the given incubation time fibrinogen was still adsorbed to the particles and thus detectable. The same principle applied to albumin. This protein is by far the most abundant one in plasma, and therefore it is most likely that it is adsorbed immediately after plasma contact. The fact that it is not very often observed on solid surfaces might be due to extremely rapid displacement within a fraction of a second. In this study a transient adsorption of albumin could be demonstrated: By diluting the plasma down to 0.08 % the concentration of the displacing species (perhaps fibrinogen) again was decreased to such an extent that albumin was still detectable after five minutes of incubation.

Assuming there were no more additional displacement steps and disregarding the minor proteins on the gels, albumin was displaced by fibrinogen, which in turn was displaced by PLS:6, apoJ and apoC-III. For the detection of a complete adsorption and displacement sequence more plasma concentrations and/or different incubation times at a low plasma concentration could be studied.

Generally for the early stages of protein adsorption - here represented by strongly diluted plasma - the hypothesis of a rapid adsorption of more abundant proteins and the subsequent displacement by less abundant ones was supported. Thus 2-D PAGE in

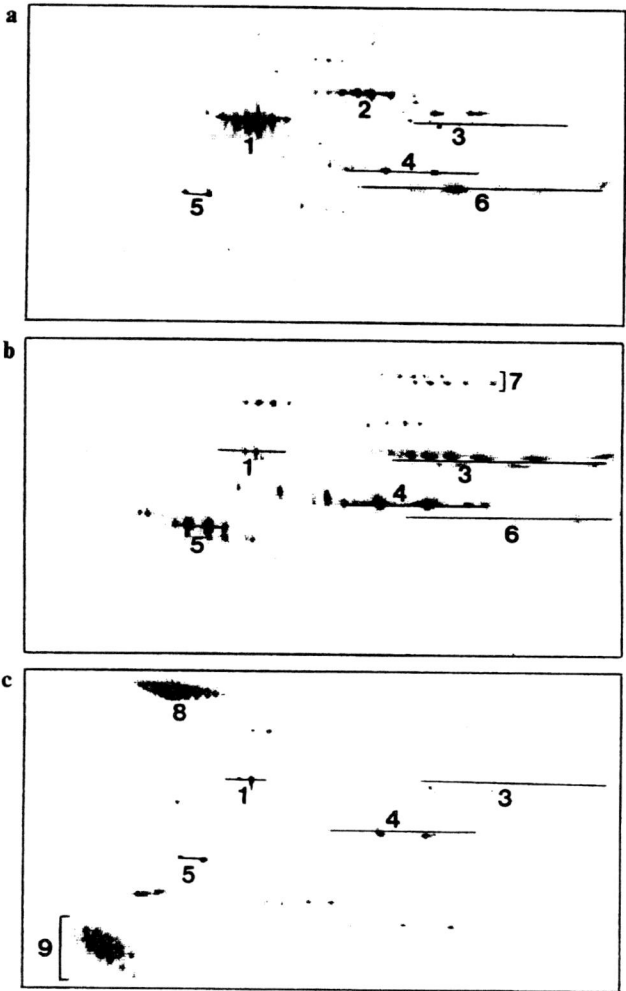

Fig. 18-3: Demonstration of the Vroman-Effect: Protein adsorption from plasma with different concentrations onto polystyrene model particles (diameter 1 μm) surface-modified with poloxamer 407 (PS 1000-407).
a) Adsorption from 0.08 % (v/v) plasma, **b)** adsorption from 0.8 % (v/v) plasma, **c)** adsorption from 80 % (v/v) plasma.
Incubation time of each sample: 5 min.
(1) Albumin, (2) Transferrin, (3) Fibrinogen α, (4) Fibrinogen β, (5) Fibrinogen γ, (6) IgG γ, (7) Plasminogen, (8) PLS:6, (9) Apolipoprotein J.
Close-ups: pI 4.8 - 8.0 (from left to right, not linear), MW 30 - 110 kDa (from bottom to top, not linear)

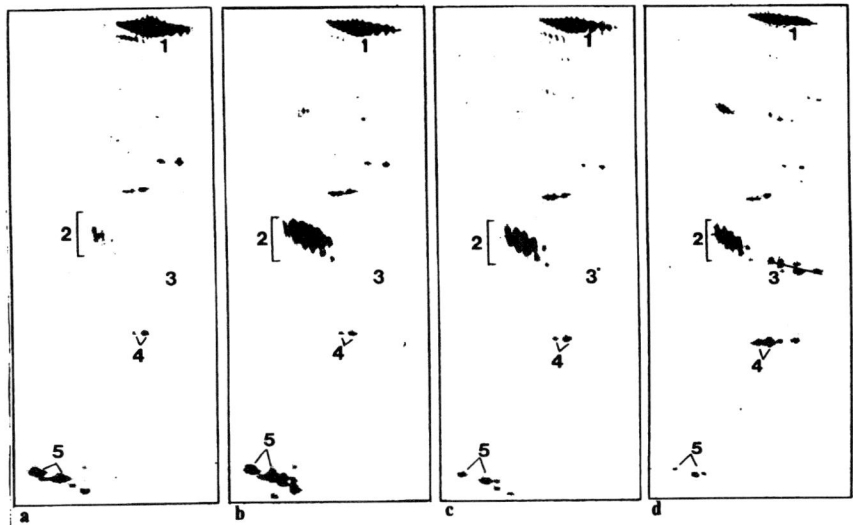

Fig. 18-4: Kinetics of protein adsorption: Protein adsorption patterns on PS 1000-407 after different incubation times.
Incubation time: **a)** 0.5 min, **b)** 5 min, **c)** 30 min, **d)** 240 min.
Plasma concentration in each sample: 80 % (v/v).
(1) PLS:6, (2) ApoJ, (3) ApoE, (4) ApoA-I, (5) ApoC-III.
Close-ups: pI 4.4 - 6.0 (from left to right, not linear), MW 6 - 110 kDa (from bottom to top, not linear), after (20).

combination with incubation in diluted plasma proved to be able to illustrate the existence of the Vroman Effect, for fibrinogen and even for albumin, in a clear and simple way.

18.3.3 Protein Adsorption within Minutes / Hours

Though protein adsorption on a time scale of seconds or even fractions of a second is interesting from a more academic point of view, more relevant to the *in vivo* fate and possible drug targeting of injected carriers is to study the adsorption and the kinetics of adsorption over a period of minutes and hours. Therefore, the protein adsorption onto the same particles as above was performed from 80 % plasma, the incubation was terminated after 0.5 min, 5 min, 30 min and 240 min.

The most important part of the gels, which contains the detected time-dependent differences in the adsorption patterns, is shown in Fig. 18-4. As compared to the gels in Fig. 18-3, the patterns presented here differed not that strongly with regard to qualitative aspects, though quantitative changes of the adsorbed protein amounts could be detected. The main proteins are PLS:6 and the apolipoproteins A-I, C-III, E and J. As the overall protein amounts on the gels differed only very slightly, in the following the amounts of the single proteins can be expressed as a percentage of the overall amount. The changes with time of the main proteins are shown graphically in Fig. 5. Over the whole period of investigation the dominating protein clearly was PLS:6, although its percentage decreased from 45 % after 30 min to 33.5 % after 4 hours. The apolipoproteins showed the biggest relative changes. ApoC-III decreased drastically, from 18.1 % after 0.5 min down to below 1 % after 4 h. ApoJ only represented 2.7 % after 0.5 min, but its amount increased strongly, so that after 5 min there was a maximum of 16 %. After that a decrease to 10.7 % was detected. Considerable increases in the amounts of apoA-I and apoE were detected. Whereas on the first three gels apoE was almost negligible (only up to 1.5 %), after 4 h it possessed a percentage of 9 %. ApoA-I increased up to a similar value, although the change was less marked.

Apart from the percentages of the single proteins, the ratios of the amounts of two or more proteins might be important for the *in vivo* distribution of the drug carriers. In the above experiment certain ratios changed drastically with time. For example, the ratio apoC-III / apoE dropped from 181 after 0.5 min down to 0.09 after 4 h, a greater than 2000-fold change. *In vivo* such changes could easily lead to alterations in the organ distribution of the particles. An example is described by Aalto-Setälä et al. [25], who reported on lines of transgenic mice which possessed an increased rate of apoC-III expression. The very low density lipoproteins (VLDL) of the mice, therefore, contained markedly increased amounts of apoC-III, as compared to the VLDL of control mice, whereas the amounts of apoE, however, were reduced. *In vivo* turnover studies and hepatoma cell culture experiments revealed a strongly diminished fractional catabolic rate of the VLDL particles in the transgenic mice indicating poor recognition by lipoprotein receptors. This change in catabolism was ascribed to both the increased apoC-III and the decreased apoE on the VLDL particles. Future *in vivo* studies, in connection with *in vitro* experiments utilizing 2-D PAGE analysis, should reveal if similar mechanisms also apply to injected drug carriers.

18.4 Conclusion

In the early stages of adsorption a rapid sequence of adsorption and displacement was found. More abundant proteins were displaced by less abundant ones. In the time frame of minutes to hours slower changes of the adsorption patterns were detected with only small qualitative alterations. However, distinct quantitative changes were detected, which led to drastic changes of certain ratios of amounts of single proteins.

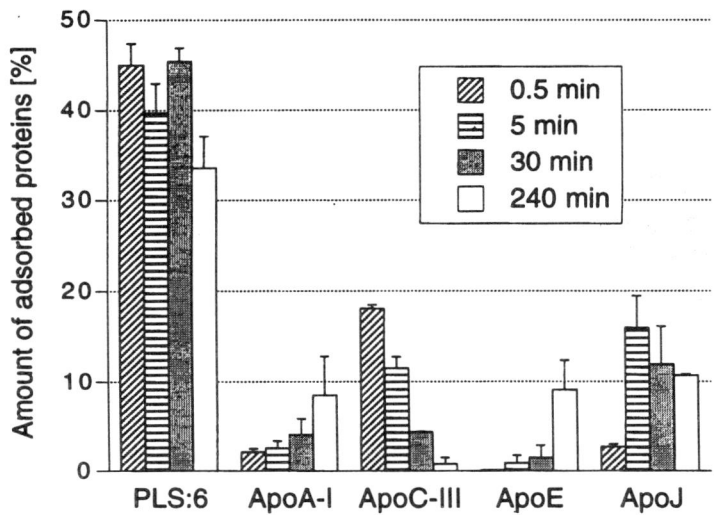

Fig. 18-5: Kinetics of adsorption of the major proteins on PS 1000-407. Amounts of single proteins are expressed as percentages of the overall adsorbed protein amounts. (Percentages represent the mean of two experiments, error bars = standard deviation).

Currently for most of the cases it is still unknown, on which principles the displacement of adsorbed proteins by others is based. Concentration in the bulk solution seems to play a major role in the very early stages of adsorption, but protein affinity for the surface (free energy of adsorption), protein-protein interactions in the layer, activation energy of adsorption [2] and possible changes in protein conformation upon adsorption may decisively influence competitive adsorption and consequently the adsorption kinetics as well. As the adsorption process is very complex, today an exact prediction of protein adsorption patterns and kinetics of protein adsorption is impossible. Therefore experiments such as those discussed above must be performed to gain further knowledge of the sequences of adsorption on particular surfaces.

18.5 References

(1) Norde, W., Adsorption of proteins from solution at the solid-liquid interface. Adv. Coll. Interf. Sci., 25 (1986) 267-340.

(2) Brash, J.L., Protein adsorption at the solid-solution interface in relation to blood-material interactions. In: Brash, J.L. and Horbett, T.A. (Eds.), Proteins at Interfaces. Am. Chem. Soc., Washington D.C., 1987, pp. 490-506.

(3) Juliano, R.L., Factors affecting the clearance kinetics and tissue distribution of liposomes, microspheres and emulsions. Adv. Drug Deliv. Rev., 2 (1988) 31-54.

(4) Dunn, M.J. and Burghes, A.H.M., High resolution two-dimensional polyacrylamide gel electrophoresis. I. Methodological procedures. Electrophoresis, 4 (1983) 97-116.

(5) Dunn, M.J. and Burghes, A.H.M., High resolution two-dimensional polyacrylamide gel electrophoresis. II. Analysis and applications. Electrophoresis, 4 (1983) 173-189.

(6) Müller, R.H. and Heinemann, S., Surface modelling of microparticles as parenteral systems with high tissue affinity. In: Gurny, R. and Junginger, H.E. (Eds.), Bioadhesion - Possibilities and Future Trends. Wissenschaftliche Verlagsgesellschaft, Stuttgart, 1989, pp. 202-213.

(7) Blunk, T., Plasmaproteinadsorption auf kolloidalen Arzneistoffträgern. Ph. D. thesis, University of Kiel (1994).

(8) O'Farrell, P.H., High resolution two-dimensonal electrophoresis of proteins. J. Biol. Chem., 250 (1975) 4007-4021.

(9) Miller, M.J., Computer-assisted analysis of two-dimensional gel electrophoretograms. In: Chrambach, A., Dunn, M.J. and Radola, B.J. (Eds.), Advances in Electrophoresis Vol. 3. VCH, New York, 1989, pp. 182-215.

(10) Appel, R.D., Hochstrasser, D.F., Funk, M., Vargas, J.R., Pellegrini, C., Muller, A.F. and Scherrer, J.-R., The MELANIE project: From a biopsy to automatic protein map interpretation by computer. Electrophoresis, 12 (1991) 722-735.

(11) Goerg, A., Postel, W. and Guenther, S., The current state of two-dimensional electrophoresis with immobilized pH gradients (A review). Electrophoresis, 9 (1988) 531-546.

(12) Hochstrasser, D.F., Harrington, M.G., Hochstrasser, A.-C., Miller, M.J. and Merril, C.R., Methods for increasing the resolution of two-dimensional protein electrophoresis. Anal. Biochem., 173 (1988) 424-435.

(13) Anderson, N.L. and Anderson, N.G., A two-dimensional gel database of human plasma proteins. Electrophoresis, 12 (1991) 883-906.

(14) Golaz, O., Hughes, G.H., Frutiger, S., Paquet, N., Bairoch, A., Pasquali, C., Sanchez, J.-C., Tissot, J.-D., Appel, R.D., Walzer, C., Balant, L. and Hochstrasser, D.F., Plasma and red blood cell protein maps: Update 1993. Electrophoresis, 14 (1993) 1223-1231.

(15) Merril, C.R., Goldman, D. and Van Keuren, M.L., Simplified silver protein detection and image enhancement methods in polyacrylamide gels. Electrophoresis, 3 (1982) 17-23.

(16) Poehling, H.-M. and Neuhoff, V., Visualization of proteins with a silver "stain": A critical analysis. Electrophoresis, 2 (1981) 141-147.

(17) Merril, C.R., Detection of proteins separated by electrophoresis. In: Chrambach, A., Dunn, M.J. and Radola, B.J. (Eds.), Advances in Electrophoresis Vol. 1. VCH, New York, 1987, pp. 111-139.

(18) Blunk, T., Hochstrasser, D.F., Sanchez, J.-C., Müller, B.W. and Müller, R.H., Colloidal carriers for intravenous drug targeting: Plasma protein adsorption patterns on surface-modified latex particles evaluated by two-dimensional polyacrylamide gel electrophoresis. Electrophoresis, 14 (1993) 1382-1387.

(19) Smith, P.K., Krohn, R.I., Hermanson, G.T., Mallia, A.K., Gartner, F.H., Provenzano, M.D., Fujimoto, E.K., Goeke, N.M., Olson, B.J. and Klenk, D.C., Measurement of protein using bicinchoninic acid. Anal. Biochem., 150 (1985) 76-85.

(20) Blunk, T., Lueck, M., Calvoer, A., Hochstrasser, D.F., Sanchez, J.-C., Mueller, B.W. and Mueller, R.H., Kinetics of plasma protein adsorption on model particles for controlled drug delivery and drug targeting. Europ. J. Pharm. Biopharm., 42 (1996) 262-268.

(21) Vroman, L., Adams, A.L., Fischer, G.C. and Munoz, P.C., Interaction of high molecular weight kininogen, factor XII, and fibrinogen in plasma at interfaces. Blood, 55 (1980) 156-159.

(22) Vroman, L. and Adams, A.L., Adsorption of proteins out of plasma and solutions in narrow spaces. J. Coll. Interf. Sci., 111 (1986) 391-402.

(23) Schmaier, A.H., Silver, L., Adams, A.L., Fischer, G.C., Munoz, P.C., Vroman, L. and Coleman, R.W., The effect of high molecular weight kininogen on surface-adsorbed fibrinogen. Thromb. Res., 33 (1983) 51-67.

(24) Breemhaar, W., Brinkman, E., Ellens, D.J., Beugeling, T. and Bantjes, A., Preferential adsorption of high density lipoprotein from blood plasma onto biomaterial surfaces. Biomaterials, 5 (1984) 269-274.

(25) Aalto-Setälä, K., Fisher, E.A., Chen, X., Chajek Shaul, T., Hayek, T., Zechner, R., Walsh, A., Ramakrishnan, R., Ginsberg, H.N. and Breslow, J.N., Mechanism of hypertriglyceridemia in human apolipoprotein (apo) CIII transgenic mice. Diminished very low density lipoprotein fractional catabolic rate associated with increased apo CIII and reduced apo E on the particles. J. Clin. Invest., 90 (1992) 1889-1900.

18.6 Abbreviations

α2-M'glob., α2-macroglobulin;
apo, apolipoprotein;
CHAPS, 3-(cholamidopropyl)dimethylammonio-1-propanesulphonate;
C, crosslinker concentration divided by the total acrylamide concentration;
2-D, two-dimensional;
DTE, dithioerythritol;
IEF, isoelectric focusing;
IgG, immunoglobulin G;
MPS, mononuclear phagocytic system;
PAGE, polyacrylamide gel electrophoresis;
PDA, piperazine diacrylamide;
SDS, sodium dodecyl sulphate;
T, total acrylamide concentration,
Vh, volt x hours,
VLDL, very low density lipoprotein.

Address of the author:
Dr. Torsten Blunk
Massachusetts Institute of Technology (MIT), E25-324
45 Carlton Street
Cambridge, MA 02139
USA

19 Index

—A—

α2-M'glob. 265; 267
α2-macroglobulin cf. α2-M'glob.
Abbé, Ernst 88; 99
Abrasive Developments Ltd 80
absorption **29**; 36
ACF 4
activated coal 202
adsorbed proteins 162
adsorbent 189
adsorption isotherm
    Rose Bengal 221
adsorption pattern 216
adsorptive 189
AFM 101; 105; 110
    constant force mode 112
    constant height mode 112
    contact mode 109
    force gradient 124
    Force-distance curves 117; 118; 119
    frictional forces 115
    laser interferometry 110
    lateral force microscopy 115
    non-contact image 126
    non-contact mode 122
    optical sensitive mode 110
    principle setup 115
    scanning force microscopy 121
    topographic imaging 109
    tunneling method 110
agethane 236
albumin 265; 267
albumin particles 215
alkyl-agarose 229
alkyl-sepharose 235
analytical method 27
    ensemble 27
    non-ensemble 27
Angular Scattering 28
Antarox™ CO 246
Antarox™ DM 246
APD 4
apo 269
apoA-I 265; 267; 272
apoA-IV 216
apoC-II 216
apoC-III 216; 269; 272
apoE 216; 272
apoJ 269; 272
apolipoprotein cf. apo
Areameter 199
Atomic Force Microscopy cf. AFM
ATR 143
    spectra 149; 150; 151
    crystal 144
    element 144; 147; 148
attenuated total reflection spectroscopy cf. ATR
autocorrelation function cf. ACF
autocorrelator 4
avalanche photodiode cf. APD
Avogadro's constant 194

—B—

B.E.T.
    concurrent dosing 197
    constant 194
    DIN norm 207
    equation 194
    Gemini-principle 210
    multi-point measurement 199
    plot 194
    practical measurements 195
    sample preparation 200; 211
    single-point measurement 199
bacteria
    hydrophobicity 215
Beer-Lambert law 65
Bessel function 9; 20; 33; 36; 58
biocompatibility 261
biological environment
    interactions 215
biomaterials 261
blood-material interaction 261
body protein 215
bone marrow
    endothelial cells 216
Brewster angle 140
Brillouin Scattering 28
Brownian motion 2
Brunauer 190; 193
BS7501 72
butyl-agarose 236; 255

—C—

cantilever 101; 109; 117; 123
capillary force 108
carboxydextran 251
catalyst K306 202
catalyst Ka 202
catalysts 185
CD 159
    adsorbed proteins 162
    chromophores 161
    electric field vectors 159
    protein monolayer 165; 167
    secondary structures 163
    spectra 160; 161; 163
Ceramide 181
cerebroside 176
Certified Reference Material cf. CRM
chaotropic effect 235
CHAPS 264
chemisorption 190
(cholamidopropyl)dimethylammonio-1-propanesulphonate, 3- cf. CHAPS
cholesterol 176
chondroitin-4-sulfate 251
circular dichroism cf. CD
CLSM 85
    applications 94
    detailed setup 89
    dyes 92; 93
    immunoliposome-cell interaction 94
    labeling procedures 94
    laser 91
    limits 92
    principle setup 87
    principles 86
    resolution 88
coal
    activated 202
coefficient
    multipole 20
collecting optics
    double lens 43
colloidal carriers 262
confocal laser scanning microscopy cf. CLSM
conformation of polypeptides 164
contact angle measurements 217
CONTIN algorithm 7; 8; 16
Cotton-effect 160
Coulomb forces 106
Coulter counter 27
Coulter® LS Series 25
Coulter® N4 Plus 2; 8; 12
Coulter® SA 3100 195
Coulter® LS 230 39; 54
counter
    Coulter 27
    optical 27
CRM 78
cumulants method 3; 6

—D—

data fitting 39
data retrieving 39
decay constant 5
delay time
    lineary 11
    logarithmically 12
    quasi-logarithmically 12
Deming 190
design qualification cf. DQ
detector array 42
dextran 251
dibucaine 179
differential adsorption 262

differential polarization microscopy 171
diffraction cf. scattering intensity
  anomalous 37
diffusion coefficient 5
Diffusive Wave Scattering 28
dimyristoyl-phosphatidylcholine (DMPC) 174
dioleoylphosphatidylcholine (DOPC) 175
dioleoylphosphatidylglycerol (DOPG) 175
diphenylhexatriene 172
distribution
  Gaussian 7
  LogNormal 7
  particle size 28
  Rosin Rammler Sperling Bennet 7
  Schulz Zimm 7
  volume-weight size 16
dithioerythritol cf. DTE
DLS 1
DOPC 175
DOPG 175
doxorubicin 179
2-D PAGE 261
  isoelectric focusing 263
  sample preparation 264
  SDS-PAGE 263
DQ 73
drug carriers 261; 262
drug targeting 262
DTE 264
dynamic light scattering DLS
dynamic sizing range 12
dynamic surface heterogeneity 170

—E—

Electric Field Scattering 28
electromagnetic wave 130
electron spin resonance cf. ESR
electron tunneling 103
electrostatic attractive force 107
Emmet 193
EN29000 71
EN45001 72
endocytosis 180; 181
entropy 229
ESR 171
ethyl-agarose 255
exponential sampling 7

—F—

FDA 69; 70
Fermi level 102
fibrinogen 265; 267; 269
fluophores
  9-anthryl-vinyl 172
fluorescence anisotropy 174; 176
fluorescence digital imaging 171
fluorescence microscopy 170
fluorophore 172
Food and Drug Administration cf. FDA

forward scattering cf. diffraction
Fourier lens 22; 26; 42
  reverse 26
Fraunhofer approximation 22
Fraunhofer diffraction 28; 32
  cf. scattering intensity
  limitations 37
Fraunhofer scattering intensity
  3-D display 37
Fraunhofer theory 36; 58
free energy 229
Fresnel diffraction 36
Frustrated Total Reflection cf. FTR
FTIR
  absorption module K 152
  scattering module S 152
  techiques 129
FTIR-ATR 143
  effective thickness 145
  penetration depth 144
FTR 143

—G—

ganglioside 176
ganglioside GM1 181
ganglioside GM3 181
gas adsorption
  carrier gas method 191
  gravimetric methods 189
  methods 191
  principles 189
  volumetric methods 189
gel-like microdomain 170
generic system validation 73
GLP 70
GM1 181
GM3 181
Good Laboratory Practice GLP
Good Manufacturing Practice GMP

—H—

Hamaker constant 107
heterodyne type 2
hexyl-agarose 236; 255
HIC 216; 229; 235
  coated polymers 241
  emulsions 247
  interactions 230
  iron oxide dispersion 251
  liposomes 248
  Mini-HIC 232; 255
  polystyrene particles 237
  theory 229
homodyne type 2
Huygen's Principle 36
hydrophobic interaction 215; 229; 235
Hydrophobic Interaction Chromatography cf. HIC
hydrophobicity 215

—I—

IEEE Software Engineering Standards 76
IEF 262; 263
IgG 216; 269

immunoglobulin G cf. IgG
indomethacin 179
infrared spectroscopy 129
  group frequencies 132
  reflection 130
  reflection spectroscopy 138
  transmission 130
  transmission spectroscopy 133
installation qualification cf. IQ
interatomic repulsive forces 107
interference 36
Internal Spectroscopy cf. IRS
International Organisation for Standardisation cf. ISO
Intralipid® 67
IQ 75
IR reflection-absorption spectroscopy cf. IRRAS
iron oxide dispersion 251
IRRAS 164
IRS 143
ISO 6; 70; 78
ISO 9000 27; 71
ISO 9001 71
ISO/IEC Guide 25 72
isoelectric focusing IEF
isotherms 192
  categories 190
  types I-VI 192

—K—

KBr press technique 136
Kramers-Kronig transformation 142
krypton 195; 209
Kubelka-Munk function 152

—L—

Laboratory of the Government Chemist cf. LGC
LALLS 75; 78; 79
Lambert's Law 134
Lambert-Beer's law 134
Laplace equation 7
Laplace inversion 7; 14; 16
laser 28; 50; 91
  argon ion 4
  diode 4
  He-Ne 4
laser diffraction 28; 57 cf. also angular light scattering intensity measurements
  calculation of volumetric concentrations 65
  concept of validation 69; 73
  data inversion 64
  energy plot 59
  ISO standard 72
  method validation 72
laser diffraction particle sizer setup 58
Lateral Force Microscopy cf. LFM
latex particles 236
  binding constants 221
  surface hydrophobicity 219
  zeta potential 219
least-squares fitting 6

Index

Legendre polynomial 20
Lenard-Jones potential 107
LFM 115
   image 117
   principle 116
LGC 210
light
   electromagnetic field 19
   infrared 29
   ultraviolet 29
light flux 43
light microscopy 85; 99; 187; 188
   epi-flourescence 85
   resolution 99
light scattering 27; 29; 54
   collective samples 22
   detector array 40; 41
   double lens optics 41
   elastic 27
   electromagnetic theory 19
   Fraunhofer 36
   geometry 31
   multiple 30
   optical model 39
   Rayleigh 30
   technologies 28
   variables 30
   wavelength dependence 30
lipid drug carriers
   effect of drug loading 178
   surface structure 177
lipid particles 169
lipid surface structure 171
lipid transfer 180
liquid-crystalline state 174
Lorenz-Mie theory 59
Low angle Laser Light Scattering
   LALLS
Low Angle Light Scattering 57

—M—

macromolecule chains 11
macropores 187
magnesium stearate 209; 212
magnetic force 108
magnetic resonance imaging cf. MRI
Mastersizer 65
matrix conversion 38
maximum entropy method 7
Maxwell equations 31
MCA 69
mean size 28; 41
Medicines Control Agency cf.
   MCA
melittin 177
mercury-intrusion 187; 188
mesopores 187
methylene-blue 189
microdomain
   gel-like 170
micropores 187
microstructure 186
Mie
   Gustav 31
Mie - factor 22
Mie approximation 23
Mie scattering cf. scattering
   intensity

   generalized 37
Mie scattering intensity
   3-D display 34; 46
Mie theory 9; 11; 19; 21; 29; 31;
   32; 65
Mini-HIC 232; 255
Minsky, Marvin 87
momentum-transfer vector K 44
mononuclear phagocytic system
   cf. MPS
motions
   rotational 5
   translational 5
MPS 261
MRI 251
multipole coefficient 20

—N—

N.I.S.T. 195
NAMAS 69; 72
National Institute for Standards
   and Technology cf. N.I.S.T.
National Measurement and
   Accreditation Service cf.
   NAMAS
National Voluntary Laboratory
   Accreditation Program cf.
   NVLAP
NBS 1003b glass beads 78
NBS 1004a glass beads 78
nitrogen 195; 209
NMR 171
   phosphorus 171
   proton 171
non-negative least-squares
   technique 16
nonylphenols
   ethoxylated 246
nuclear magnetic resonance cf.
   NMR

—O—

ocean, color 30
octyl-agarose 236
operational qualification cf. OQ
opsonin 215
optical model 41
optical rotatory dispersion cf. ORD
OQ 75
ORD 159
organ
   artificial 261

—P—

P828 178
palmitoylsphingomyelin 174
particle characterization
   resolution 40
   submicron particle 43
   trimodal distribution 53
Pauli exclusion principle 106
PC 174
PCS 1; 28
   channel arrangement 11; 13
   error sources 14

Fingerprint method 14
matrix format of ACF 16
multiangle 4
multiangle measurement 9; 14
signal-to-noise ratio 13
PE 174
PEM 166
PEM-IRRAS 166
pentyl-agarose 236
performance qualification PQ
permeametry 187; 188
phagocytes 215
phagocytosis 219
Phase Doppler Analysis 28
phosphatidylcholine (PC) 174
   anthrylvinyl-labeled 175
phosphatidylethanolamine (PE)
phosphatidylglycerol 175
phosphatidylserine 176
phospholipase A2 169
phospholipids
   anthrylvinyl-labeled 172; 174
photoelastic modulator cf. PEM
photomultiplier tube cf. PMT
photon correlation spectroscopy
   cf. PCS
Photon Migration Measurement 28
physisorption 190
PIDS 23; 43; 45; 47; 48; 53; 61
   data analysis 51
   oblate spheroid 62
   optical setup 50
   signals 25
   single non-absorbing spheres 61
   system 24
   trimodal distributions 52
piezoelectric ceramic 114; 123
pigment powder 66
piperazine diacrylamide cf. PDA
plasma protein 216
   preferential adsorption 216
PLS:6 265; 267; 269
PMT 4
Polarization Intensity Differential
   Scattering cf. PIDS
polarization scattering 47
Poloxamer 216; 241; 243
Poloxamine 908 241
polyacrylamide gel 263
polydispersity index 14
poly-l-lysine 162; 163
polymer adsorption 215
polystyrene latex
   Rose Bengal partitioning 223
polystyrene latex particles 3; 9;
   14; 137; 149; 236; 266; 269
   cf. PSL
polystyrene particles
   fluorescent 222
pores
   analytical methods 187
   classification 187
   gasadsorption techniques 189
   shapes 186
PQ 75
procaine 179
propanolol 179
propyl-agarose 236
protein adsorption 261

kinetics 268
protein B 265; 267
protein secondary structures 163
proteinkinase C 178
proteoliposome 177
Provencher 7
PSL 52; 53

—Q—

Q spec process 74
QELS 1
Qspec Validation Package 71
qualification cf. validation
Quality Assurance 73
quasi-elastic light scattering cf. QELS

—R—

Raman Scattering 28
Rayleigh approximation 21
Reference reticle 79
reflection spectroscopy 138
   Attenuated Total Reflection cf. ATR
   diffuse reflection 151
   external reflection 138
reflection-absorption spectroscopy cf. IRRAS
refractive index 5; 30; 35; 63
   conformation of correctness 65
   determination 64
regularized non-negative least-squares 7
Regulatory Compliance 69
Regulatory Enforcement Agency 69
RES
   clearance 216; 217
Rose Bengal 216; 217
   adsorption isotherms 218; 220; 221
   binding constants 221
   binding methods 218
   partition quotient PQ 223
   partitioning method 223
   structure 220

—S—

Salmonella typhimurium 395 216
salting out effect 235
Scanning Electron Microscopy cf. SEM
Scanning Force Microscopy cf. SFM
Scanning Force Spectroscopy cf. SFS
Scanning Near-Field Optical Microscopy cf. SNOM
Scanning Probe Microscopy cf. SPM

Scanning Tunneling Microscopy cf. STM
Scatchard plot 219; 221
scattering intensity 32; 36
   angular pattern 32
   fine structure 41
   Fraunhofer 36
   non-spherical particle 36
   polarization effect 35; 45
   polystyrene latex 33
   refractive index effect 36
   wavelength effect 35
Scattering Under a Flow Field 28
scattering vector 5
SDS 263
SDS-PAGE 262; 263
SDS-polyacrylamide gel electrophoresis cf. SDS PAGE
self-beating mode 2
SEM 101
sepharose CL-4B 236
SFM 101
SFS 121
   image 122
Siegert relation 5
sieving 42
silver-staining technique 262
singular value analysis 7; 16
size distribution 28
sky, color 30
SLN 113; 117; 119; 120; 126
SM 174
SNOM 102
sodium dodecyl sulfate cf. SDS
software cf. validation
solid lipid nanoparticles cf. SLN
SOP 71; 75
specification qualification SQ
sphingomyelin (SM) 174; 176
SPM 101
SQ 73
Standard Operating Procedure cf. SOP
static surface heterogeneity 170
STM 101; 102
Stokes-Einstein equation 5
Ströhlein 199
surface area
   determination 185
surface heterogeneity 170
surface hydrophobicity 215; 229
   latex particles 219
   scale 223
surface roughness 187

—T—

Teller 190; 193
TEM 100
tetracaine 179
TickIT 76
tissue
   artificial 261
transferrin 269
Transient Scattering 28

Transmission Electron Microscopy cf. TEM
transmission spectroscopy 133
transmittance T 135
Triton X-100 231; 235
tungsten lamp 50
tungsten-halogen lamp 50
tunneling 102
tunneling current 103
Turbidity Measurement 28
Two-Dimensional-Polyacrylamide Gel Electrophoresis cf. 2-D PAGE

—V—

validation 73
   analytical light energy distribution 81
   change control 77
   design qualification 73
   generic system 73
   installation qualification 75
   non-validated software 80
   operational qualification 75
   performance qualification 75
   primary verification 78
   secondary verification 79
   software 76; 80
   software environment 76
   software verification 82
   specification qualification 73
   verification "kit" 82
   verification of LALLS 78
van der Waals forces 106; 107; 190
verification cf. validation
VLDL 272
"V" life cycle 76
volume-to-intensity conversion factor 16
voxel 85
Vroman, Leo 268
Vroman-Effect 268

—W—

WELAC 72
Western European Laboratory Accreditation Co-operation cf. WELAC

—X—

xenon 195
X-ray diffraction 187

—Y—

yellowish green 222

—Z—

Zeolon 500 202
zeta potential 219
zinc oxide 202